THE FLOWERING OF THE PACIFIC

The Flowering Of The Pacific

Brian Adams

Being an account of
Joseph Banks' travels in the
South Seas and the
story of his *Florilegium*

COLLINS
BRITISH MUSEUM
(NATURAL HISTORY)

Designed by Patrick Coyle

© Brian Adams 1986

First published in 1986 by William Collins Pty Ltd, Sydney
Typeset by Setrite Typesetters, Hong Kong
Printed by South China Printing Co, Hong Kong

The National Library of Australia
Cataloguing-in-Publication data:
Adams, Brian, 1934 –
The flowering of the Pacific

Includes index.
ISBN 0 00 217472 3.

1. Banks, Sir Joseph, 1743 – 1820.   2. Cook, James, 1728 – 1779.
3. Banks' Florilegium.   4. Botany — Pacific Area.
5. Endeavour (Ship).   I. British Museum (Natural History).
II. Title.
581.99

CONTENTS

COLOUR PLATES FROM
BANKS' FLORILEGIUM

Plates appear in the order of printing and issue by Alecto Historical Editions.

For Loraine

PREFACE

Two hundred years is little more than a moment for ancient civilizations, but it embraces most of the recorded history of a vast area of the earth's surface: the South Pacific. In the mid-eighteenth century, following the end of their Seven Years War, Britain and France were in an expansive mood, eyeing distant territories not yet formally claimed. The world as viewed from northern Europe still seemed to offer far-flung trading prizes because the largely unexplored southern hemisphere was thought to contain a great continent somewhere between the known land masses of South America and New Holland. The vision of a country peopled by noble savages, living in harmony with nature and their gods amid bounteous riches, began as fantasy, turned to conjecture and then became a fixation. All that was needed to take advantage of this great southern continent, savants thought, was to be quick off the mark, make the certain discovery and appropriate it before rivals raised their flags of annexation.

That was the background against which the *Endeavour* bark of the Royal Navy set sail from England in 1768 bound for the other side of the globe, ostensibly on a scientific mission to study an infrequent planetary phenomenon, but secretly in search of new dominions. The time was propitious for investigating both the secrets of the heavens and the produce of the earth because the art of navigation based on astronomic principles was in its infancy, and the idea of studying plant species for their own sake was brand new, with botany just emerging from its herbal role as an adjunct to medicine. Lieutenant James Cook, the *Endeavour*'s bluff Yorkshire captain, knew about navigation, but his gentlemen passengers, led by the mercurial young Joseph Banks, the self-styled 'botanizer', mystified him. However, it was to be a voyage on which botanists might make their name in this Age of Enlightenment, whose intellectual obsession had become the nature of Nature, and it would set the pattern for future scientific missions.

What was discovered after the vicissitudes of three years' voyaging in unknown seas was surprisingly little according to the terms of reference. No new territories could be claimed, as Tahiti, New Zealand and Australia, though not called by those names, were already known to exist, and, at the exhausting end of their travels, Cook considered it necessary to make a grovelling apology to his masters in London, taunting himself for failure to present them with major disclosures. The great southern continent remained a geographical phantom to haunt the Georgian mind. In fact, the Lords of the Admiralty were delighted with the immaculate surveys which fixed the position and delineated outlines of New Zealand and the east coast of New Holland for the first time. James Cook's prodigious maritime skills eventually would be recognized as central to the *Endeavour* voyage being one of the greatest achievements in the history of navigation and exploration.

Joseph Banks and the survivors of his retinue took back with them a treasure-trove of botanical drawings and specimens that would form the first record of South Pacific flora and, together with the many other examples of natural history collected or recorded, greatly expand knowledge of the region. Additionally, the observations of Australasian geography would lead within 20 years to the establishment of a colony, whose existence was directly linked to the visit of the sturdy little *Endeavour* and her motley crew.

History treats its heroes indiscriminately and, despite concerted efforts at the beginning of this century to have Sir Joseph Banks certified as 'the Father of Australia', it was the austere, self-contained Cook who captured public imagination and achieved considerable acclaim by monument, although he had demonstrated a lack of any further interest in the continent after ignoring it on two subsequent voyages to the Pacific. No statuary was erected to the influential Joseph Banks in the land he recommended for colonization and, furthermore, the great botanic publication based on the copper plate line engravings he commissioned was never published in his lifetime, making it one of the mysteries and minor tragedies of eighteenth-century science.

The printing of the entire edition of what is now known as *Banks' Florilegium* has generated a revival of interest in the man and his work. The project, due for completion in 1988, was undertaken by Alecto Historical Editions and the British Museum (Natural History). It has had, altogether, a gestation period of some 25 years: three on the voyage itself, 12 when the plates were being meticulously engraved at a cost in today's money of more than £750,000, and, 200 years later, a decade devoted to its printing. There has never been a series of line engravings like these 738 exquisite botanical portraits, produced in a limited edition of 100 sets.

The following is an account of Joseph Banks' and James Cook's voyage into the unknown and the final realization of its scientific fruits by an equally courageous modern printing enterprise. The narrative begins with the voyage of the *Endeavour* to her main landfalls at Madeira, Rio de Janeiro, Tierra del Fuego, Tahiti, New Zealand, Australia, Java and home by way of the Cape of Good Hope. It progresses in chronological sequence through Banks' increasingly busy and celebrated life as the pivotal figure of British science for 40 years, the attempts to get the engravings published after his death, and the re-kindling of interest in the project during the 1960s.

A bibliography at the end of the book lists principal publications about Banks, Cook and other members of the ship's complement. The appendix includes technical and distribution details of the *Florilegium*, and chapter notes state the main sources of information, together with additional information about the *Endeavour* voyage, and eighteenth-century botanical illustration and production procedures for *Banks' Florilegium*.

PART ONE
THE VOYAGE

1
London — Madeira
1768

*No man will be a sailor who has contrivance
enough to get himself into a jail, for being
in a ship is being in a jail, with the chance
of being drowned.*

Samuel Johnson (1709–1784)

IN the late summer of 1768 work at the Plymouth dockyard of the Royal Navy was being carried out in a relaxed mood under warm sunshine tempered by refreshing breezes from the south-west. The yard had been extended some years earlier to service the Western Squadron in time of war and saw intense activity during the seven years' conflict, when naval blockades of France's ports and those of her ally, Spain, were sent out from Plymouth. That bitterness and bloodshed had ceased five years before and they were back to the peacetime routines of repairing and building the king's warships. A keen observer standing on the grassy expanse of the Hoe and looking down on the maritime activity below, might have noticed something unusual about the ships in the harbour. There was the predictable scattering of sloops, frigates and men-o'-war, but in their midst was a type of vessel rarely seen in these waters. With uncompromizingly bulky construction and displaying no figurehead or other decorative flourishes, she looked like a square-rigged coastal trader, painted dull sulphur yellow on her upperworks with a band of black pitch all around and brown at the waterline. However, naval colours were displayed aloft from her new spars and fresh rigging, and there were too many men on the crowded deck to be merely for the purpose of carrying cargo.

The 368 ton vessel was arriving for final preparation before leaving England on a special mission to the other side of the world. Although she resembled a North Sea collier, blown far from her normal trade routes along the busy sealanes of the east coast, these appearances were deceptive and would lead to some unfortunate misunderstandings before very long. The *Endeavour* bark, under the command of Lieutenant James Cook, was virtually a new ship, from stem to stern, from false keel to topmast, fitted out to exacting specifications so she might survive the stresses and strains of negotiating distant and uncharted waters. This lumbering former coal-carrier from Whitby had been transformed over the past few months at Deptford yard into an efficient machine for discovery. With shallow draught

The *Earl of Pembroke* (later *Endeavour*) leaving Whitby Harbour. Painting by Thomas Luny.

and breadth of beam she could be taken into shoal waters and get herself off sandbanks, an ideal configuration for exploration in unknown seas.

They anchored in the sound and then moved into the dockyard for several days of refitting, which included, according to her captain's journal, 'completing the gentlemen's cabin'. Extra stores were taken aboard and four carriage guns were delivered at the dock to add to her armaments. Three additional seamen signed on, 12 marines reported for duty, the whole company was read the Riot Act and the conditions of service on board ships of His Majesty's Navy and then paid wages for two months in advance. There would be no need for conventional currency where they were going and the petty officers and seamen made arrangements for allotments to families and parents during their absence.

Meanwhile in London one of the *Endeavour's* gentlemen passengers was saying his farewells to friends, family and sweetheart. Joseph Banks, a wealthy landowner from East Anglia, owner of a fine town residence in New Burlington Street, fashionably-dressed, handsome, 25 years old and Fellow of the Royal Society, was about to take a coach bound for Plymouth to board the *Endeavour* and set out on his latest adventure. He had been accepted as a paying passenger on an official scientific voyage to the South Seas to observe the transit of Venus across the face of the sun. He and his companion intended to collect plant specimens and document such aspects of living nature as caught their fancy. Banks was not very interested in geography or the science of astronomy because he was not yet given to deep philosophical thought or even scientific notions. But this was the Age of Enlightenment and men like him were thirsty for knowledge of the world around them. Well-endowed young gentlemen were expected to take the grand tour to Paris, Venice, Florence and Rome to look at classical antiquities and broaden their minds before returning home to the comforts of

Georgian England for a life of leisure in whatever fashion they chose. Often it led to debauchery and gaming; to Banks, however, that was for block-heads. He needed to go where few had ventured before and so his grand tour would be around the world.

Joseph Banks' passion for botany began during his schooldays and had led him to the Lee and Kennedy nursery garden beside the Thames at Hammersmith. The nursery was run by a Scottish Quaker named James Lee, whose business among the landed gentry and aristocracy flourished to the point where his selection of rare flowers and plants rivalled those of the royal gardens at Kew. One of Lee's other responsibilities was a beautiful young ward, Harriet Blosset, who became infatuated with the dashing young Joseph Banks, who, it was known, displayed a keen interest in pretty young women as well as horticulture and botany. Banks' reputation in this regard was notorious and the flirtatious Harriet soon became one of his conquests. Although he claimed to have no interest in settling down to a domestic life while there was still a whole world to explore, Banks was often seen together with Harriet at dinner parties and the theatre. The man was in love and finally agreed to marry his sweetheart after he returned from his trip.

A ring was purchased and Harriet promised to spend the time during his absence living in rural seclusion embroidering waistcoasts for her fiancé. They spent what would prove to be his last night in London holding hands during a performance of *La Buona Figliuola* by Niccolò Piccinni at the Haymarket Theatre on Friday, 15 August. It was a wildly popular *opera buffa* whose story about young love triumphant suited the mood of the evening. Banks was excited about the voyage ahead, Miss Blosset melancholy at the prospect of being parted from her lover for an indefinite time but, as in the opera's plot, sure of his return and their eventual bliss. After the final curtain there was a supper party for friends and family. Banks drank heavily to hide mixed feelings about not having been able to bring himself to tell Harriet that he would be gone by morning. The next day, too early for tearful farewells, he was to be off on the long journey to Plymouth, responding to a letter from James Cook summoning him on board the *Endeavour*.

There was plenty of time to think about the future as the coach carrying Banks and his companion, Dr Daniel Solander, who had taken leave of absence from his job at the British Museum, left the relatively smooth highway out of the metropolis and began a lurching journey over the rural tracks that passed for roads. The jolt of leaving Harriet subsided in the throbbing of a hangover and Banks decided he was not at all sorry to be leaving the rigid patronage-controlled society of Britain and the small establishment group to which he belonged — the pleasantly ordered, yet claustrophobic nucleus of privilege. He could also forget the worries of his estates for a while, with their demands for supervision of crop management and agricultural economics. In intellectual matters, all aspects of knowledge were being pursued: agriculture, medicine and science, and Banks was eager to be seeking out new dimensions and discoveries in the expanding world. At this moment of boneshaking inconvenience a long sea voyage into limitless adventure seemed attractive, even therapeutic, and it was easy to overlook that, all too soon, the random rocking of the coach would give way to another type of uncomfortable motion where privacy was much harder to come by. For the forseeable future his home would extend in one direction less than the length of two cricket pitches and in the other, a mere ten easy paces. Into this space, barely three times longer than it was broad, would have to fit 85 officers, seamen and marines, together with an astronomer and his servant and Banks' own retinue of seven — a total of 96 men in a vessel that, as a collier, was handled by four able seamen and a clutch of apprentices. Nine years before, the acerbic conversationalist Samuel Johnson observed that sailing was tantamount to being jailed. He had added: 'A man in jail has more room, better food and commonly better company'. For the young botanist only the first would ring true.

Earlier in the year Banks had successfully lobbied the Royal Society, Britain's most prestigious scientific body, to include him on its proposed astronomical expedition to the South Seas to observe the transit of Venus. A precise measurement of this infrequent phenomenon taken from various points on the globe could fix the distance between earth and the sun and, as a standard unit of measurement, would lead to improved techniques of navigation. The first transit to be observed was by a young English curate with a great gift for astronomy, Jeremiah Horrocks, in 1639, although his results were not published until many years after his death. It was left to the astronomer Dr Edmond Halley to point out in a paper read to the Royal Society in 1691 that by the accurate observation of this phenomenon the distance of the sun from the earth might be determined with the greatest certainty. Observations of the first transit after the publication of his paper were made in 1761, but they were not successful; some calculations were incorrect and many of the best locations could not be used because of the war between Britain and France. The 1769 transit was of vital interest because there would not be another one for more than a century.

While the French, Danes, Swedes and Russians all planned their own viewing stations, the Royal Society considered King George the Third's Island (later named Tahiti), recently discovered by Captain Samuel Wallis on the naval ship *Dolphin*, an ideal site for one of the observations, although British astronomers would also go to Spitzbergen and Hudson's Bay. James Cook, aged 39, a brilliant navigator, surveyor and chartmaker was put in charge of the expedition, which had the blessing of the king and some financial aid from the royal coffers. An ambitious young Scotsman named Alexander Dalrymple had hoped to be appointed leader because of his expert knowledge of Pacific exploration gained through years of study while working for the East India Company. Dalrymple had corresponded with the French expert on South Seas voyages, Charles de Brosses, and with his assistance brought up to date the information on the subject, which would be published in his book *Voyages in the South Pacific Ocean*. Alexander Dalrymple had been the Royal Society's choice as leader of the proposed scientific voyage, but as the expedition was to be undertaken by a warship, regulations stated that its commander must be a naval officer. After much acrimony from Dalrymple who considered, with justification, that he was the most suitable person for the job, the command was handed to Lieutenant James Cook, complying with regulations, and because there was to be more than just astronomical observation.

Cook's secret instructions would also include a search for the great southern continent, an elusive territory that had intrigued European minds for two centuries, adding political overtones to what was publicly a scientific journey. The expeditions sent out to the South Seas during the peace with France after 1763 began a new era in exploration. Previously, the results of voyages into the unknown, whether privately sponsored or financed by the state, tended to be suppressed for commercial and political advantage. The secrecy surrounding these journeys to the Pacific during the sixteenth and seventeenth centuries resulted inevitably in extravagant rumours and gross speculation about great land masses, colourful inhabitants and vivid rituals. The explorations of Cook's immediate predecessors were different because they were carried out with political motives. Now, in the late eighteenth century, the discovery and annexation of new territories would continue to add to the power of both Britain and France but, at the same time, many of the scientific details of the voyages were intended for publication.

On the basis of the Royal Society's request and the king's patronage, the Admiralty had decided to buy a ship specially for the expedition. In choosing the *Endeavour*, they took the type of craft Cook knew well. She was originally named the *Earl of Pembroke*, technically a cat-built bark, only three-and-a-half years old, a sturdy type of cargo vessel. The son of a Yorkshire day labourer, James Cook grew up on these cats before transferring to a different kind of life in the Royal Navy.

After a steady but unspectacular career, which had seen him serve on blockades against the French during the Seven Years War and participate in several bloody sea battles in the Channel, first as an able seaman and then as a master, James Cook was transferred to the North American station during the tense times between Britain and France in Canada. In 1759, shortly after the fall of Quebec and the end of French sovereignty there, he was appointed to the *Northumberland*, from which he was to survey the St Lawrence River, which he did until 1762.

The publication in 1763 by the Admiralty of his charts and sailing directions for the area brought Cook's name to prominence within his profession. Further good work in Labrador and Newfoundland gained him promotion to first lieutenant and led to the captaincy of the *Endeavour* at a salary of five shillings a day.

Cook hoisted his pennant on taking over his first major command on 25 May 1768, but it was not until two months later that he was officially informed of the extra passengers. These included Mr Banks, who was described by the Lords of the Admiralty as 'a gentleman of large fortune, well versed in natural history'. Cook was instructed to receive an additional party of eight, together with their baggage and to victual them 'as the bark's company during their continuance on board'.

The *Endeavour's* purchase price had been £2,800 and a further £2,294 was spent in repairs, alterations and special equipment. This included sheathing the hull with a second skin of planking to protect her timbers against the voracious teredo worm of the South Seas. Banks himself expended almost this total again to finance his part of the voyage, although it was going to be far from a comfortable grand tour. Cook was dismayed that so many extra people were to be crammed into his tiny ship but he could not protest; the Admiralty's instructions had to be followed implicitly and however inconvenient it would be to share his quarters and conveniences with the gentlemen passengers, he had to accede.

James Cook was a commanding figure, just over six feet tall, large-boned with strong features that some regarded as good looking. His quick and piercing eyes under prominent brows set him apart from his fellow officers, although a modified Yorkshire accent and sensitivity gave the impression of a modest person. He was also temperate, stubborn, not given to religion or politics, and liked his food. Above all, he was clearly in command.

Cook and Banks had met formally only a few weeks previously at the Deptford naval dockyard on the Thames, where the *Endeavour* was being prepared for her long journey. The meetings were pleasant and correct, as befitted a rich gentleman adventurer of influence and a career officer of humble background whose paths had crossed only once before — briefly — in Newfoundland. The young botanist used his wealth to ensure that his was better fitted out for the study of natural history than any previous expedition had been. This meant bringing to the ship a confusing variety of stores and equipment for Cook's officers to arrange safe stowage — nets, trawls and drags, bottles for preserving small animals and fish in spirit, chemicals to treat plants and seeds, drawing materials, paper and paints. He could not resist taking along a few expensive toys as well, including a device based on the idea of the telescope for looking at the bottom of the sea in clear water, and two electrical machines. Experiments with electricity were very popular at this time in philosophical circles, although Banks would use his apparatus mainly for playing practical jokes on unsuspecting victims. There was also an extensive portable library containing mainly reference books on natural history, and special delicacies to ease the monotony of naval food rations. Banks ordered his own supplies of the best salt beef, salted cabbage, brandy, porter and small ale; he also arranged to have on board live sheep, pigs and poultry together with his two pet greyhounds. As the ship would take some time to make her way round the coast to Plymouth and then have to wait for extra work to be completed, it was decided that Banks should join the *Endeavour* when Cook was ready to

Portrait of James Cook by Nathaniel Dance, c1772.

depart. Other members of the Banks party together with the official astronomer, his servant and everyone's baggage went aboard at Deptford.

As the coach lurched along south-westwards on its three-day journey
Banks remembered the previous time he had taken this route, more than
two years before. He told Solander about the thrill of setting off in the
spring of 1766 for the unlikely destinations of Labrador and Newfoundland.
That journey had come about in an unusual way. His mother was a deeply
religious woman and, after being widowed and going to live in Chelsea, had
come into contact with the Moravian Church. The followers, who practised
a type of Presbyterian worship and were also known as the United Brethren,
took their gospel to the Eskimos on the coast of Labrador, where Britain
had interests in the rich cod fishing grounds (cod was a valuable supply of
food and lamp oil, made from the livers). Banks was fascinated to hear the
stories told by Moravians back from Labrador, of the animals, Eskimos and
plants, and he was determined to go and see for himself. He wished to
become part of that stream of British seamen, explorers, hunters, adventurers, writers and scientists who had been visiting the nation's first over-
seas colony since it came under British jurisdiction by the Treaty of Utrecht
in 1713. In normal circumstances it would have been impossible to entertain
such a wild fancy, but social connections helped fulfill most of his wishes.
A friend from his schooldays was able to pull strings. The result was that
Banks could sign on as supernumerary in a naval vessel sent for the protection of British fisheries against incursions by the French. She was HMS
Niger, under the command of Sir Thomas Adams, which left Plymouth
bound for the north-west in late April 1766 with Banks' fishing gear, plant
presses, butterfly nets and notebooks safely stowed away in the hold.

The journey helped to prepare the young man for his forthcoming trip to
the other side of the world and banished any illusions about the glamour of
sea travel. It was an uncomfortable expedition, though it brought great
satisfaction to Banks, who was the first to collect plant specimens systematically in the region. He arrived home at the end of 1766 to a relieved
mother and sister.

During his absence Banks was elected to the Royal Society on nomi-
nations that included the Bishop of Carlisle and the librarian of the British
Museum. There had been some grumbles about wealth and privilege aiding
his selection and a few queries about scientific seriousness, but any doubts
were soon dispelled when the collections from Newfoundland and Labrador
were revealed, although there was to be no formal account of the expedition.
Banks was not particularly eager to have his name connected with the
discoveries, an attitude that stemmed from his status as a gentleman amateur,
setting the pattern for future collecting and documentation.

The 23-year old heir to the Revesby Abbey agricultural estates near
Boston in Lincolnshire returned to London with his appetite whetted for
more adventure. Even if his background and education hardly fitted him for
either, he was developing a taste for serious travel and study in natural
history. Banks' parents had planned a social advancement for their only son,
seeing his future as something better than the normal life of country landed
gentry. For him, it had been Harrow at the age of nine and then Eton at 13,
where he did not shine in Greek or Latin and was unenthusiastic about
games. But he made valuable contacts and his form master described him as
'a very good-tempered and well disposed boy'.

Every country house at the time had a copy of John Gerard's *The*

Herball or Historie of Plants, first published in 1598. It was an essential reference to good husbandry in the garden and also contributed, through its illustrated plates, to embroidery designs for the ladies. Mrs Banks at Revesby was no exception in possessing the *Herball* and it triggered in her son an interest in plants that was sustained by rambles through the lanes near college and along the banks of the Thames until natural history became an obsession. Eton satisfied his parents' social aspirations by association with the English establishment. This was consolidated when Banks matriculated at Oxford in 1760 and became a gentleman commoner at Christ Church. The college offered a distinct social cachet: one student had written a few years earlier that it gave the opportunity 'to make any acquaintance that may be useful in future life', adding, with candour, it was 'the only reason that I am sent to this college'. Universities in England, however, were at a low ebb at that time with poor attendances and, lacking the vital flow of vigorous young intellects, they became withdrawn from the centre of affairs. Tutors and lecturers were receiving slightly better pay from the church or the college and no longer needed to impress potential students in order to earn a living by giving private lessons. The standard of accommodation had deteriorated and bad reports filtered back to noble families, who often directed their sons to alternative finishing education, such as the European grand tour. The Age of Enlightenment and the Newtonian revolution largely passed Oxford by and in Banks' admittance year there were 200 freshmen, of whom only ten came from the aristocracy.

The Professor of Botany was Humphrey Sibthorp, but in his many years of occupying the chair, he was remembered as having given only one lecture in the university that boasted the oldest botanical garden in England. There was no teaching available in the subject and Banks himself organized the services of a 'proper person', one Israel Lyons from Cambridge University, the son of a Jewish watchmaker with an interest in astronomy as well as botany. When Banks was on vacation and staying at his mother's house in the village-like atmosphere of Chelsea, he also learned about practical and theoretical botany from Philip Miller, who had been in charge of the adjacent Chelsea Physic Garden since 1722. He was the author of the very popular *Gardener's Dictionary* and friend of the great Swedish botanist Linnaeus, who described him as a prince of botanists as well as a prince of gardeners. A neighbour at Chelsea, which had its own society, was John Montagu, the fourth Earl of Sandwich. In his early forties, a musician, passionate fisherman and, at various times, Lord of the Admiralty, Montagu had estates in Lincolnshire where he and Banks often fished, well-endowed with food and drink as well as rod and line.

Banks persevered at Oxford and left with an honorary degree, unconcerned about academic achievements. Most of his fellow students had regarded the act of attendance at university as a springboard to public careers in politics or private lives as country gentlemen. He came into his inheritance in early 1761 and could enjoy financial security for life.

Cook had planned to leave Plymouth by the third week of August and was ready to depart on the twentieth, but the winds blew steadily from the south-west, which was an unfavourable quarter for a square-rigged ship unable to sail close to the wind, and he had to wait for moderating conditions. This delay gave the passengers an opportunity to become familiar with their new surroundings and get to know one another. Banks had chosen his own party carefully, with an eye to expert skills backed by faithful service.

10

His companion and fellow gentleman, Daniel Solander, was a 33-year old Swede, a graduate of Uppsala University and former pupil of the great Carl von Linné, who had become one of Banks' heroes. Von Linné (who assumed the Latin version of his name — Linnaeus) reformed the science of botany with his system of classification, generated by a need to create order in natural history, and was greatly gifted in his ability to communicate his love of the natural world to his pupils and disciples. Solander had arrived in England nine years before to study gardens and museums and his own obsession with botany resulted in a desire to classify, according to the Linnaean system, every living plant on the face of the globe. Banks was still a student when they first met at the British Museum where Solander, already a Fellow of the Royal Society, secured a position of assistant librarian. He helped Banks with information for his Newfoundland expedition, and their mutual devotion to botany led to a firm friendship. Soon after Banks had been granted permission to sail with the *Endeavour*, he and Solander were dining together at Lady Monson's, discussing the journey ahead and what might be discovered. The normally taciturn Solander, fired with enthusiasm for what he was hearing, suddenly jumped up from the table and asked, 'Would you like a travelling companion?' Banks replied without hesitation, 'A companion like you would be of infinite advantage and pleasure to me'. 'In that case I will go with you', was the reply, and next day Banks applied to Sir Edward Hawke, the First Lord of the Admirality, for permission to take Solander. At first he was refused, but youthful charm was able to win over the obstacles of officialdom.

The main botanical specimens collected on the voyage would be drawn by Sydney Parkinson, the son of a Quaker brewer from Edinburgh and great admirer of the quintessential artist of Georgian England, William Hogarth. When his father's business collapsed Parkinson had become apprenticed as a wool draper, a job that eventually took him to London. He had already shown talent as a draughtsman and attracted the attention, through Quaker connections, of James Lee at Hammersmith, who introduced him to Joseph Banks. Parkinson was commissioned by Banks to copy a collection of drawings brought back from Ceylon and then sent to work on living specimens at Kew. The excellent results lead to his appointment as botanical draughtsman on the *Endeavour* at the age of 23. The other artist was Alexander Buchan, also Scottish, a young man of talent given to epileptic fits, although that would not be revealed until later. His main task was to draw the coasts, the landscapes, and also the natives they would encounter. Another member of the party was Herman Diedrich Spöring, aged about 38, who came from Finland, where his father was a professor of medicine at the University of Abô. Spöring studied surgery in Stockholm and then moved to London, where he became a keen naturalist. He was a good draughtsman and Banks engaged him as a personal secretary. The four remaining members of the entourage were servants: two came from the Revesby Abbey estates — James Roberts was a boy of 16 and Peter Briscoe had accompanied his master on the Newfoundland and Labrador journey; the others were negroes named Thomas Richmond and George Dorlton, who added a touch of contemporary elegance, a continual reminder of London society where it was the height of fashion to have black servants. The other gentleman passenger was the astronomer Charles Green, who was charged with the responsibility of supervising the observations of the transit; he was aged 33, a Yorkshireman, like Cook. Green had led a chequered career in astronomy,

starting at Greenwich Observatory as assistant to the Astronomer-Royal in 1761 and continuing there for four years. In 1763 he was appointed with Dr Nevil Maskelyne to the commissioners of the Board of Longitude for determining the best way of fixing longitude. Inexact methods for this greatly hindered finding a ship's precise position at sea. When Maskelyne was made Astronomer-Royal in 1765, Green left Greenwich and engaged in private astronomy until he was appointed by the Royal Society to the present mission.

On 25 August, the wind shifted to NNW and at four in the afternoon the *Endeavour* was able to set sail, aided by what Cook called 'a nice little breeze off the land'. As the *Endeavour* butted her way past Drake's Island and faced the open sea, it was a time for reflection on when they would see England again. For many it was a matter of *if* they would return, since high death tolls were not uncommon on such journeys. Cook knew from the recent voyages through the Pacific and around the world by Byron and Wallis on the *Dolphin* that the duration could not be less than two years; Banks assumed an absence of three. Cook thought they set out in a cheerful mood and with a 'readiness to prosecute the voyage'. Banks confirmed this by admitting that they were all 'perfectly prepared (in mind at least) to undergo with cheerfulness any fatigues or dangers that may occur'. The botanist John Ellis wrote to Linnaeus in Sweden: 'No people ever went to sea better fitted out for the purpose of natural history'. The books of the great Swede were on board, essential for the systematic classification of plants. The naturalist, Gilbert White, summed up the spirit of the venture when, in writing about Banks, he stated: 'May we hope that this strong impulse, which urges forward this distinguished naturalist to brave the intemperance of every climate; may also lead him to the discovery of something highly beneficial to mankind? If he survives, with what delight shall we peruse his journals, his fame, his flora?'

The gentle breezes held constant and the heavily-laden vessel, which was quite obviously no greyhound of the sea, sailed before them at her maximum speed of a plodding six or seven knots; as much as could be expected from such blunt bows and chunky build. The flat bottom caused consider-

Left: Portrait of Mrs Sarah Banks by John Russell, 1779.

Centre: The young Joseph Banks. School of Gainsborough. Probably painted as a retrospective portrait after the *Endeavour*'s return.

Above: Sydney Parkinson self-portrait, c1768.

Opposite: Portrait of Joseph Banks by Joshua Reynolds, 1775.

able movement even in a slight sea and Banks was furious with himself at feeling increasingly queasy, finding the ship 'a heavy sailer' but resigned to suffer the inconveniences of her construction 'which is much more calculated for storage than for sailing'. Buchan and Parkinson busied themselves preparing sketching materials and the ship's company settled down to their routines of shipboard life. Then the wind turned to a westerly gale and seasickness got the better of most passengers, with the motion so violent that Parkinson was unable to hold pencil to paper. For Banks it was not quite as bad as his outward trip on the *Niger* two years before, when he had had to rope himself to a gun on deck to avoid being swept overboard, so weakened was he by continual retching and unable to face going below. The old saying, 'A man who went to sea for pleasure would be likely to go to hell for a pastime', began to ring true. This time he lay propped on his cot in the claustrophobic cabin, reeking of vomit, listening to the creaking of timbers, the loud slapping of the rigging and the roar of water along the keel, untended by his servants, who were similarly afflicted. This rough weather across the Bay of Biscay carried away one of the *Endeavour*'s small boats, which had been lashed to the deck, and washed overboard pens containing several dozen live poultry, brought on board to provide the gentlemen with fresh eggs for breakfast. They would not have had a stomach for omelettes until the storms passed, but it was a sad loss because laying hens could not easily be replaced at the planned ports of call. In the meantime Banks prostrated himself, yearning for the nausea to pass as he was forced to listen to what he remembered Homer calling 'the loud-resounding sea', seemingly amplified by the construction of this bouncing bark.

By September they were on an African latitude out in the Atlantic and the bouts of seasickness subsided as the winds lightened, allowing Banks to drink a toast to Europe astern and adventures ahead. Cook's plan was to head for the Portuguese island of Madeira to take on wine and fresh provisions before picking up the trade winds that would carry them across the equator and south to Rio de Janeiro on the coast of the Brazils. Afterwards, it would be due south with perhaps a call at the Falkland Islands or Tierra del Fuego before running the gauntlet of the notoriously changeable weather around Cape Horn. That safely negotiated, the *Endeavour* could head for her appointment with Venus far out in the immensity of the Pacific.

Banks had regained his sea legs after getting used to the idiosyncratic movements of a cat-built bark and was eager to see some nature study under way. He embarked on what he described as his 'first essay on the inhabitants of the sea' by catching and examining jellyfish, or 'blubbers' as the seamen called them. Some species of plankton were netted, carefully noted by Solander and then drawn by Parkinson, who could now work with paper and pencil again. Specimens were preserved in spirit jars and Banks was happy that the scientific part of the voyage had begun.

On the evening of 12 September the *Endeavour* came to anchor at Madeira in 22 fathoms just off the port of Funchal. All day they had been beating along the coast of this jagged island, a huge mass of rock thrust up from the Atlantic ocean, but now the lights of the capital glinted across the bay to welcome them. It was too late to move to a closer anchorage, so the night was spent in the stream close to the British naval ship *Rose* and several merchant vessels. At first light the next morning the anchor was weighed but both winds and tide were against the *Endeavour* and she was almost

14

impossible to manoeuvre in such changeable conditions. At first, instead of moving towards the shore they found themselves drifting away from their destination. This must have confused the Portuguese soldiers manning the batteries because the commandant of one of them assumed the vessel was leaving Madeira without reporting to the authorities and two shots were fired across her bows as a warning. This caused momentary consternation on board because Funchal was regarded as a friendly port and had long been a regular calling place for the British to take on the island's famous wine. Cook ordered his men to ignore the threat and was able to bring his vessel round to the shore and anchor in 15 fathoms, making a point not to salute the fortifications on the way, which was traditional for visitors. The British consul, Mr Cheap, was horrified to learn what had happened and quickly informed the governor of the incident, expressing official disappointment that a ship of the British navy should suffer such indignity. An apology was received with the assurance that the officer responsible for the unwelcome reception would personally apologize, if it was desired. Cook declined the offer, dismissing the incident as a minor affair.

They arrived at an island that legend held was discovered by an Englishman, Robert Machin, who was shipwrecked on this uninhabited land with his mistress in 1346. Documented history gave the honour of discovery to the Portuguese explorer João Gonçalves Zarco, who claimed it for Prince Henry the Navigator in 1420 and gave it the name *A ilha de madeira* — wooded island, because of the dense forests. For more than a century-and-a-half Madeira grew and prospered under Portuguese rule and because of its location became a port of call for ships travelling to the east. Later it became a stepping-stone to the New World, one of its most famous visitors being Christopher Columbus, who married the daughter of an island governor and lived for a while on the adjacent island of Porto Santo. In 1580, together with all of Portugal, it came under Spanish control until an uprising on the mainland resulted in the country regaining independence. In the mid-1600s Charles II of England married Catherine of Braganza and their union resulted in advantages for the British merchants on the island, particularly those involved in the wine trade, which flourished with vines brought from the rich volcanic soil of Crete and Cyprus. The early settlers had soon discovered that fruit and vegetables would grow in profusion with the combination of fertile land and a benevolent climate, and Madeira became famous for its fresh produce. When the *Endeavour* arrived, wine was the main product of the island and its rather acid quality, derived from the volcanic soil, needed blending and fermentation before stabilizing. Brandy was added, the process giving a touch of sweetness most obvious in the malmsey, dark brown in colour, with a rich smoky tang and the sharpness on the palate common to all madeira wine. It became the staple drink of the early British settlers in America and then delighted many palates in England.

The gentlemen looked across from the *Endeavour* to the impressive panorama before them. The town of about 8,000 inhabitants was situated within a bay and Banks thought it looked larger than the island deserved. The waterfront fortifications, governor's palace and the fine fifteenth century cathedral were set against sharply rising serrated mountain peaks. Cultivation of the soil seemed intense, right up to the highest points, with vines and citrus trees in profusion, causing Parkinson to remark, 'It looks like one, wide, extended garden'. There was a strong resident British trading

Portrait of Joseph Banks by Francis Cotes, c1768.

community with a consul, vice-consul and 22 merchants who had their own hospital. In the track of commerce had come rich British invalids to recuperate in the gentle, flower-perfumed climate when, according to a contemporary report, 'their native air fails to produce relief'. Banks and Solander were invited to live ashore during the brief stay as guests of the consul, Mr Cheap. They were anxious to travel out of the town as much as possible to spend time in the country collecting plants, although it was not an ideal time of year for botanizing, with the vintage in full swing and the countryside brown. Banks could see in the distance plantations of bananas, guavas, pineapples, all products of a more gentle climate than he had known before, and he was eager to get out and examine them. Horses and guides were made ready but representations to the governor on their behalf for permission to inspect the nearby countryside took a whole day, after which it was learned they would be allowed to travel only in twos. A hasty explanation from Mr Cheap that his friends wished only to inspect the island's flora gained them greater liberty to go about their researches, but they were limited to a radius of about three miles from the town because of the brevity of the stay. Banks and Solander used most of the daylight hours collecting and, in spite of the season, many species of plants as well as insects were gathered, together with several varieties of fish between journeys to the hills. There was plenty to keep them busy, sorting, classifying and recording, preparing for the time when working routines would be established across the long stretch of ocean after leaving Madeira.

While the gentlemen were making what they called their 'excursions into the country' work continued on the *Endeavour*, including maintenance of decks, spars and rigging, taking on 3000 gallons of water, purchasing vegetables and acquiring a live bullock weighing more than 600 pounds. He was hauled up on the longboat's davits and added to the ship's menagerie of living animals, which included sheep, pigs and goats. In the midst of these regular operations there was a first taste of death when Alexander Weir, a master's mate who had been standing in one of the boats alongside the *Endeavour* supervising the slipping of an anchor, caught his legs in the buoy rope, was thrown overboard and taken to the bottom by the anchor's weight. By the time they were able to raise him to the surface, he had drowned. His death was the first of many that could only be expected on such a voyage and it elicited little concern from Weir's fellow seamen. Sydney Parkinson, however, confided a few private thoughts to his journal and described him as 'a very honest worthy man and one of our best seamen'.

Apart from accidents and misadventures, the most common cause of death at sea at this time was scurvy, often accounting for the demise of half a ship's complement on a long journey. If left unchecked, scurvy, due to a diet deficient in ascorbic acid, led to fatigue and depression, livid haemorrhaging sores on gums and skin, swelling of the joints and, inevitably, death. Cook's instructions from the Admirality included specific antiscorbutic measures and he intended to carry them out when supplies of fresh foods were exhausted. Madeira was able to provide plentiful fruit and vegetables together with fresh meat, although Banks' discriminating palate thought the local pork, mutton and beef were inferior to that produced on his own estates in England. Some of the seamen, however, had developed an addiction to salt meat and when two of them refused to eat the Funchal fresh beef they were found guilty of insubordinaton as their refusal might have incited others to follow their example. Cook committed the offenders, able seaman Harry Stephens and marine Thomas Dunster, to a standard punishment of 12 lashes each as a warning that discipline must be maintained on board. And so in beautiful Funchal Harbour, with the sun glinting on the calm blue water against the impressive backdrop of town and mountains, Stephens and Dunster were stripped to the waist and tied to a grating on the deck while a contingent of marines in uniform was drawn up around them under the command of sergeant John Edgecumbe. The young drummer boy, Thomas Rossiter, was directed to beat a tattoo and the boatswain's mate, John Reading, administered the lash to its tempo until the offenders' backs rose in ugly red weals and then bled from the onslaught of leather. Cook had made his point about diet for the moment and, to emphasize his intentions, the next day he issued the whole company with 20 pounds of Madeira onions each. It was an unusual gesture because the navy's Victualling Board in London, which set regulations for the supplies to all its ships, did not usually allow such extras and Cook would have to seek approval for the unauthorised expenditure at the end of the voyage.

On shore, the gentlemen made the most of their time collecting plant, fish and shell specimens and observing the local customs, including the curious method of transport for goods and humans in wooden sleds pulled by oxen because the streets, as Banks noted, were narrow and 'uncommonly ill paved'. Parkinson thought Funchal a rather expensive place as most commodities, including food, utensils and clothes were imported from England or other parts of Europe. Able seaman Matra also looked at the town with an enquiring eye because he had decided to compile his own journal of the voyage with the subversive intention of publishing it later. He noted that there were two hospitals, one of them exclusively for lepers, a large Franciscan college and many inelegant churches. He was impressed with the wine and sought out details of its trade, noting the grape varieties and discovering that best madeira sold for £26 a pipe; inferior wine from the north side of the island, which was drunk as *vin ordinaire* by the locals, fetched only half that price. About a third of the entire output went to Britain and her colonies and no less than six ships a year transported wine to the Portuguese possessions in Brazil, where the *Endeavour* was headed. The purchases for the ship were of the finer quality and would be improved by the motion of the journey and the high temperatures of the tropics. Almost all of Madeira's own consumption was sent to sea several times before being ready to drink. Banks thought the winemaking techniques extremely crude and probably the way Noah would have made his wine 'when he had planted the first vineyard after the general destruction of mankind and their arts'. On second thoughts, Banks considered he might have employed somewhat better methods.

Cook was kept busy on board the *Endeavour* for most of the time in port attending to details of administration. He needed to replace the dead master's mate and the standard way of doing this was by impressment, which meant stealing a sailor from a merchantman. The practice was legally based on the royal prerogative of calling on all men for military service and could extend to any seaman, although certain categories such as apprentices, fishermen, watermen and those younger than 18 or older than 59 were supposed to be exempt and could obtain certificates of protection. A party of seamen was sent out from the *Endeavour* under the command of an officer and they brought back John Thurman, who belonged to a New York sloop flying the British flag in Funchal harbour. He did not qualify for protection against the press gang and seemed to be little disturbed about his sudden transfer and altered destination.

Banks and Solander received generous assistance from Thomas Heberden, a long-time resident of Madeira and the Canary Islands, who was the chief physician of the region and practised at Funchal and Tenerife. He had been a Fellow of the Royal Society since 1761, having sent meteorological and geological information to London. He also made an extensive study of the trees of Madeira, which were notable for great stands of chestnuts and pines reaching high into the hills, and he gave a copy of a paper he wrote on the subject to Banks together with specimens and advice on where to collect interesting plants. Solander was of the opinion that Dr Heberden had more influence on Madeira than the governor and dubbed him 'the oracle of the island'. Parkinson started on his watercolour drawings, including a fine representation of one of the island's evergreen trees known locally as *aderno* with yellowish-green flowers and purple fruit. Its wood was said to be used in shipbuilding and for making wine casks. It was later classified as *Heberdenia bahamensis* after the helpful resident doctor. There was also an attractive date plum, *Diospyros lotus*, with its yellow fruit, found all around the Mediterranean, and a local herb native to Madeira and the Canaries with yellow flowers and brown seed pods called *trevina* locally, classified as *Lotus glaucus*. These plants and many more kept them all fully occupied.

There was an introduction from Dr Heberden to the nunnery of Santa Clara with its glazed tile belltower reflecting the sun and well-stocked walled garden overlooking the town. The nuns had been told that the visitors were members of the Royal Society and they were greeted as great philosophers, assumed to be able to predict all manner of natural phenomena with almost supernatural knowledge. 'With their tongues going at an uncommonly nimble rate', according to Banks, the ladies wanted to know when it would rain and where best to search for spring water in their garden. The gentlemen left for the relative peace outside the nunnery, relieved of the constant chattering that had assaulted them for the best part of 30 minutes. Young James Mario Matra, who was in the party, found the sisterhood exceedingly plain and noted, 'I did not observe one who could pretend to more than a moderate share of beauty'. Banks looked down on Funchal bay from the heights and observed chauvinistically to Solander that the climate was so fine, 'any man might wish it was in his power to live here under the benefit of English laws and liberty'.

The governor, who did not receive foreigners, announced he would visit the gentlemen. Although it meant losing a day when they might have been out collecting, protocol demanded that the travellers were present for this 'unsought honour of an official visit', as Banks described it. He decided to play a trick on his excellency by telling him of the scientific instruments carried on board, hoping that he might want to see one of them. It worked, and one of the electrometers designed by the celebrated instrument-maker Jesse Ramsden was summoned from the ship and set up in Mr Cheap's house. The governor was eager to be entertained by Banks' 'electrical and other philosophical experiments' and was shown how to hold the electrodes and receive a current. The glass wheels of the machine were turned and Solander, supressing a laugh, noted that they worked 'prodigiously well' as the dignitary jumped in surprise at the strength of the shock, muttering Portuguese oaths in a most undignified manner. Banks felt he had been revenged for the loss of his valuable collecting time.

They came to the end of their stay of five days with a collection of 330 plants thought to be native to the island and 69 others that were introduced. These were by no means the first English botanists to work in Madeira; Sir Hans Sloane collected there in the early years of the century and James Harlow botanized extensively for Sloane and Sir Arthur Rawdon. By the time of the *Endeavour*'s visit, plants from the island were well established in several gardens and herbariums in Britain. Sydney Parkinson was in fact given only 21 specimens to draw from what, after all, had essentially been a provisioning port of call and, of those, 11 plates would be engraved for the *Florilegium*. Eventually they would form the smallest section of the work. James Cook despatched a short letter to Mr Stephens, the Secretary of the Admiralty. It ended, 'having taken on board as much wine as the ship can conveniently stow, and completed our water, shall put to sea again tomorrow'.

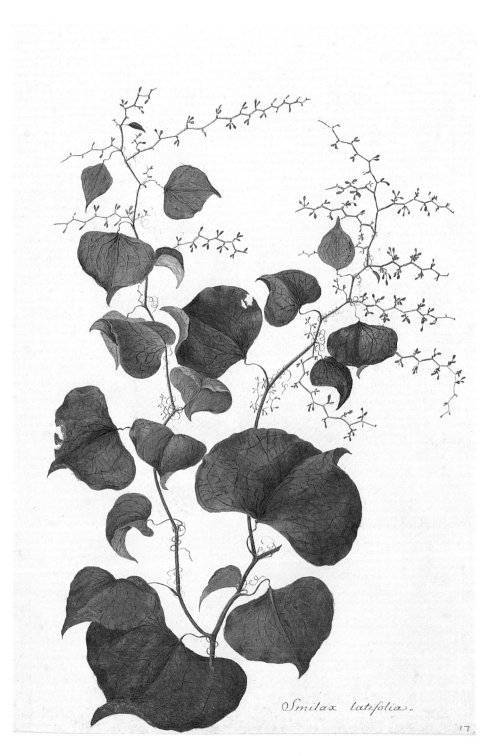

Smilax latifolia.

17.

Smilax aspera Watercolour drawing
from the *Endeavour* voyage, made at
Madeira.

2
Madeira — Rio de Janeiro
1768

On such a full sea are we now afloat,
And we must take the current where it serves,
Or lose our ventures.

William Shakespeare (1564–1616)
Julius Caesar

THE *Endeavour* sailed from Madeira in a light breeze on the evening of 18 September in as fine a condition as any vessel could be for a two-month voyage to the Brazils. There would be fresh meat to eat for the next week because the steer taken on board at Funchal had now been slaughtered and there were plenty of vegetables to accompany the boiled beef. James Cook had ordered more onions to be purchased and the day after sailing another ten pounds were distributed to each member of the crew. The decks were as clean as penned livestock allowed, the sails dry and the rigging unchafed as the ship headed for the north-east trades that would carry her south-west, across the equator, to the South American coast. Flying fish were seen skipping over the deep blue water, some landing on the deck, their sides glinting in the sunshine like burnished silver. Porpoises played about the ship, but they could not be caught.

During the first few days, the captain spent considerable time with the astronomer Charles Green working on techniques of navigation and entering the results of their calculations in the ship's log. Fixing longitude meant a series of complicated sextant sightings of the moon, the sun or various fixed stars and then referring to the Astronomer-Royal's *Nautical Tables*, published the year before. These provided the figures for observations at Greenwich and a comparison had to be made between the vessel's reading and the fixed longitude in the tables. It was a tiresome and time-consuming procedure, made necessary because of the lack of an accurate chronometer. However, in spite of the hours it could take to make a single calculation, Cook was confident that with the tables they could find the *Endeavour*'s position to within a degree and a half, and usually much less. Green was surprised to learn that none of the ship's officers was familiar with the latest means of taking observations and even Cook showed ignorance of the most efficient methods of fixing his vessel's position. The astronomer had found a satisfactory way to keep himself occupied all the way to the South Seas where his main task lay.

Joseph Banks was now able to institute regular work procedures for his staff that would lead to a comprehensive scientific record of the voyage. The normal day started at 8 a.m., when they gathered around the green baize table in the great cabin, which was a misnomer because the space available was very constricted, but at least there was enough headroom for easy movement and excellent light from the five stern windows to make conditions agreeable for exacting scientific tasks. As Banks described it: 'Seldom was there a storm strong enough to break up our normal study time'. The only interruption was dinner, taken in the middle of the day with the captain and his senior officers. The gentlemen worked at the table again from about 4 p.m., when the cabin had lost the smell of food, and sat until dark 'with our draughtsman opposite and showed him in what way to make our drawings, and ourselves made rapid descriptions of all the details of natural history while our specimens were still fresh'. The information was copied by Spöring into a journal and plants were pressed between sheets of Addison's commentary on Milton's *Paradise Lost*, a job lot of paper acquired from a London printer. It was a fine coincidence that botany and poetry contrived to combine in a phrase such as 'flowers worthy of paradise' as the *Endeavour* negotiated Milton's 'rising world of waters dark and deep'. Animal specimens, with no apparent literary connections, were preserved in glass jars supplied by a London surgeon.

When not at work in the great cabin, Banks spent every moment of daylight on deck trying to catch fish in scoops or small nets and observing bird life. One day a young shark took a bait and was hauled aboard. A quick study was made of it on deck and its four sucking fish were put aside for preservation. Banks, ever keen to experience new tastes, decided to add white shark to his list of gastronomic conquests and sent it to the cook, John Thompson, to dress and stew for dinner. The result was most acceptable to the gentlemen's palates but the seamen could not share their enthusiasm. Banks assumed with justification it was because of 'some prejudice founded on the species sometimes feeding on human flesh'. Birds visited the ship frequently. Swallows dropped on the deck in exhaustion, other species were blasted out of the sky by Banks' musket and he made a pet of a little yellow wagtail, although it was a short-lived association because the bird soon fell prey to one of the ship's cats. When calms were encountered, the young naturalist went out in his own small boat to collect surface creatures, including marine snails and jellyfish. It was a stimulating time for the gentlemen, living in an atmosphere of harmony with the captain and his officers, and a busy period for Banks' servants keeping his cabin and those of the staff clean, trimming candles, tending the dogs and helping to prepare food. The captain and his rich passenger were reasonably accommodated but the overcrowded vessel forced the officers and scientific staff to share minute cabins or canvas-sided cubicles, stacked beneath the great cabin down to the bilges.

Banks found most aspects of life at sea very enjoyable; he had provided well for himself and his retinue by way of provisions and equipment but he missed the joys of music and being able to expend the physical energy that was part of his youthful vigour. When the *Endeavour* was under way, with no opportunity of going out in the boats, there was little space for healthy exercise. He tried to solve this by gymnastic experiments with a couple of ropes in his cabin, but the little box, barely six feet square, proved to be too limited. When the ship lurched in the middle of a skipping exercise he fell, struck his head, and suffered sickness in the stomach for several hours afterwards. He decided to abandon such ideas until they reached the next port, when collecting excursions would provide an outlet for bridled exertions. The seamen below had no such concerns; leisure time for them was smoking, drinking and talking in the suffocating fug of the between-decks space that served as combined living and sleeping quarters for 40 men. Their daily duties up and down the rigging or attending to the dozens of other labours on deck that were routine matters on a long voyage kept them fit enough. In fact, the only open air space available to them was cluttered with the impedimenta of exploration. The longboat, for carrying about 22 people, took up almost a fifth of the ship's length and the pinnace, for 14 passengers, covered another 14 feet. They were securely lashed down when the *Endeavour* was under way. On either side of the main and foremast were the four hand pumps and the forward deck contained pens for pigs and sheep. The ship's principal goat, who was already a seasoned traveller, being familiar with the West Indies and having gone round the world with Captain Wallis on the *Dolphin*, was allowed to find the most comfortable spot on deck between the spare masts, which were carried to meet any emergencies aloft. The space was even more constricted towards the blunt bows with heavy balks of timber used for securing the chains when the vessel was at anchor, and there was also a selection of half-a-dozen anchors of different sizes to meet all possible mooring situations. The chimney from the cook's galley stove below smoked away invitingly from amidships, looking as if the *Endeavour* was anticipating the age of steam. In spite of all the extra equipment the ship was still recognizable in profile as a former Whitby collier, but below, down through the two hatches that led to the crew's quarters, there were many alterations. A lower deck had been installed along the entire length, dividing the former hold horizontally to provide living space above and stowage below. In the hold, also running the full length of the ship, and with a height of about 11 feet amidships, were all the areas of supply and preparation for food and drink. Coal was stored there for the kitchen oven together with casks of water used for cooking, occasional washing and mixing with rum and brandy to make grog. The stern structure, which was quite rudimentary on a working cat-built bark, had been extended forward and contained cabins on two levels above the lower deck, including the great cabin aft, where the gentlemen and officers ate, worked and relaxed. There was also the captain's quarters, together with those of Banks, Solander and Green.

It was calculated that the *Endeavour* would pass the equator on 25 October and organizing the traditional ceremony of crossing the line brought some light relief to the monotony of life at sea. Most of the company had not ventured into the southern hemisphere before and were required to be ceremonially dunked in the ocean. A list was drawn up of all living creatures on board, including the dogs and cats — but not the veteran goat and the innumerable rats — and sent to the great cabin at dinner time. An accompanying signed petition from the ship's seamen sought to know who had not crossed the line before. Afterwards everyone was called to the quarter deck and examined by John Gore, one of the lieutenants, who had made two complete circumnavigations of the globe. He marked on his list who was to undergo the ceremony, but Banks was able to evade immersion for himself, his staff and dogs by paying a forfeit in brandy while Cook excused himself in rum. Some of the ship's company, however, chose the ceremony rather than give up four days' allowance of wine, which had been fixed as the going rate for seamen, and there were 21 left who agreed to undergo this baptism, including all the boys on board.

A block was made fast to the end of the main yard with a rope through it attached to a rudimentary seat built from three lengths of timber: one for sitting on, another for holding and the third to guard against the victim's head hitting the yard when he was pulled up from the water. The boatswain gave the order for commencement by blowing his whistle, the man was hauled up as high as possible suspended over the sea, and then the rope was released sending him plummeting. This was repeated twice to fulfill the initiation. It was a lively diversion for the onlookers, who crowded every inch of the *Endeavour*'s limited deck space, laughing and cheering at each ducking. The participants, however, found it rather daunting, if not downright frightening, because few could swim and falling off the seat would lead to almost certain drowning. Most of them participated in the ceremony with a spirit of bravado, grinning and shouting as they fell to the warm ocean, but a few almost suffocated from fear.

Just south of the equator Banks thought it was time to tinker with his electrical machines again and one of them was set up in the great cabin for testing. It could be coaxed to produce only a small charge of current and the shocks were very week. Many of the components seemed to be damp, but even after drying, the results were no better. Banks, who claimed he was trying to carry out a serious experiment, blamed the atmospheric conditions for the failure. The Ramsden equipment had worked perfectly well in London and on the governor at Madeira. Charles Green had an electrometer of a different manufacture on board and when a comparative test was made that day, it worked reasonably well. The machines were then put back in their boxes and forgotten for a long time. Reading was a less frustrating way of obtaining amusement. With the books carried by Cook, Banks, Solander and Parkinson, the *Endeavour* was something of a floating library. There was the usual collection of navigational volumes on board, but the demands of the voyage for numerous scientific references and theories of Pacific exploration made it a very varied collection. Prominent in Banks' list of about 40 volumes was de Brosses' *Histoire des Navigations* and several other books referred to by de Brosses which influenced Banks' thinking about the existence of a great southern continent. Among the many botanical, geographical, zoological and ornithological works was Thomas Blanckley's *A Naval Expositor* which, with its engraved illustrations and explanatory test, enabled the gentlemen to learn about naval procedures and shipboard life, even if they disapproved of most of them. There was also an advance copy of Alexander Dalrymple's book on Pacific exploration, which was not published until after the *Endeavour* left England because the author was waiting for the return of Bougainville from his voyage around the world. All of these works, some of which were also referred to by Cook, related directly to aspects of the expedition and the type of information needed on such a long journey. Sydney Parkinson, on the other hand, revealed a cultivated background and different enthusiasms by his own book list, which included Hogarth's *Analysis of Beauty* from 1753, Homer's *Iliad* and *Odyssey*, Ovid's *Metamorphoses*, Shakespeare and *Don Quixote*, together with various histories and gazeteers.

The air was pleasantly warm during the day and at night, the temperatures quite moderate and within a narrow range, but the humidity, which had obviously affected Banks' electrometer, was extremely high. He complained about his razors rusting and black leather portfolios becoming white with mould. The servants had to wipe down many of Banks' possessions each morning, particularly the books. The close atmosphere did not seem to affect the seamen, who were forced to spend most of their leisure time below decks. Only half the crew was off-duty at any given time. The others would be standing watch of four hours on and four hours off, three times a day in rough weather. Hammocks were rolled up during the day allowing the men to sit at rough benches to eat, drink, smoke, chew tobacco and talk in what Banks described as 'the dreadful energy of their language'. Aft of the seamen's quarters, separated by a strong wooden bulkhead, were the cabins of the gentlemen and their staff and the officers. Here the atmosphere was vastly different from that surrounding 'the common people', as the young botanist called them without intentional derision. His own four servants messed and quartered with the seamen and were regarded as crew in most respects, including standing watch. The main difference between life on the *Endeavour* and that aboard a conventional warship was the accommodation for the captain and his senior officers. Being in command, Cook would normally have had spacious quarters, but Mr Banks and his party had upset the traditional arrangements and the captain, on his first major command, was forced to forgo this naval privilege for the sake of science. This was accepted by the patient and understanding Cook, although he was still mystified by the almost indecent enthusiams of the gentlemen for collecting. He could identify much more easily with his fellow Yorkshireman Charles Green, practitioner of empiricism and speaking a language of navigation in a common accent that both of them could understand.

Above the great cabin stood the quarter deck, reserved for the officers and gentlemen, where Banks strode like a caged animal, a few paces each way, in an attempt to get some exercise after the dismal attempts in the privacy of his cabin. A false deck had been constructed over the tiller on the poop, giving a little more room for the gentlemen's promenades, enabling them to walk without tripping over the tiller ropes, which normally would have swung across the space.

Cook was a stickler for discipline and hygiene, regarding the two as inseparable. As well as an assault on scurvy, he insisted on keeping the *Endeavour* clean and healthy. Banks and Solander would not work in the great cabin during the afternoons while the smell of dinner lingered in the air, and that was in an area that could easily be ventilated by opening the stern windows. Below decks in the sailors' quarters the combination of food smells, body odours, tobacco smoke and foetid bilge water was worsened by the dripping humidity of the tropics, with mould taking hold everywhere. Canvas wind scoops rigged on the hatches forced some fresh air below but not enough to create a healthy circulation. Cook had small fires lit in metal pots and taken into the bowels of the ship so that their heat could stir up the fug and the smoke fumigate areas where no draughts could penetrate.

October passed and the *Endeavour* continued her slow progress towards the coast of South America. Cook announced one day at the dinner table, 'I have now determined to put into Rio de Janeiro in preference to any other port in Brazil or the Falkland Islands'. His reasons were based on the plentiful supplies of food known to exist there and the facilities for servicing the vessel. He added, 'From the reception former ships have met with there,

I doubt not but we shall all be well received'. Banks was particularly keen to visit Port Egmont in the Falkland Islands for collecting and he assumed that might still happen, as their course down the coast of South America after leaving Rio de Janeiro was yet to be determined. On 8 November, with a smudge of land on the horizon, they made first contact with the new continent by encountering a Portuguese line-fishing boat crewed by two whites and nine blacks. Banks and Solander went on board to negotiate a price for the catch, including bream and dolphin, which was enough to feed the *Endeavour*'s entire complement. Surprisingly, the captain insisted on being paid in British shillings, and he got them — 19 shillings and sixpence. A few of the fish were kept aside for study but the bulk of the purchase provided a welcome change of diet as the fresh provisions taken on board at Funchal were exhausted. Cook had ordered sauerkraut and portable soup to be included in the men's diet from early October and madeira wine was served instead of beer, which had almost run out. Everyone relished the prospect of going ashore and tasting the abundance promised in the flourishing colonial outpost of Portugal. Parkinson had completed 100 drawings on various subjects and he looked forward to making many more in the days ahead. There was a high degree of expectation as they cruised down the coast and entered Rio de Janeiro harbour in light airs on Sunday, 13 November. Everyone was overwhelmed by the beauty of the setting. There were towering mountains in the background and the closer peak of the famous Sugar Loaf on the port bow was flanked by expanses of golden sandy beaches. Alexander Buchan drew profiles of the coastline as they passed.

The first European sighting of Brazilian territory took place in 1500 by a Spaniard, but in the same year a Portuguese navigator, Pedro Cabral, on a journey to India, landed on the coast and took possession of it in the name of his king. The harbour of Rio de Janeiro was established two years later and became a regular calling place for ships on their way to the east. Its settlement developed as a tribute to the genius of Portuguese seamanship, exploration and colonization. The outposts needed to be defended against incursions from the French as well as the Spanish, Dutch and British, who all had their eyes on strategic positions as ports of call en route to both the Pacific and Indian Oceans. When the *Endeavour* arrived in its harbour, Rio had been the capital of the Portuguese territories in South America for only five years, taking over from Bahia far to the north. It was ruled from Lisbon, which thrived on the rich products of the interior, including sugar — a European luxury — coffee and timber. Gold and precious stones came from the hinterland for the royal treasury. During the seventeenth century, in a period of huge expansion, more than 500,000 slaves had been brought from Africa and blacks greatly outnumbered the white colonists. From being a fortified outpost on the rim of the jungle, Rio de Janeiro developed a unique and elegant lifestyle, but under the administration of the viceroys it was virtually a closed port to foreign merchants. There was intense suspicion about any outside influence, particularly at this time when Portugal and Spain were wrangling over the possession of Uruguay. With its glorious harbour and fine climate, Rio boasted good livestock, a great variety of fruits and vegetables and excellent seafood, all of which made it a highly desirable place to live for the Portuguese and the other European nationalities who worked for them.

While becalmed at the mouth of the river which gave the city its name,

Cook sent Lieutenant Hicks and chief mate Richard Pickersgill in the pinnace to the viceroy to obtain a pilot. Meanwhile, a breeze sprang up and the *Endeavour* was able to proceed on her own, aided by signals from the forts on the shore. Britain and Portugal had been allies for most of the century and there was no reason to suppose that the ship's company would not be made welcome here, as in Madeira six weeks before. Cook knew that Commodore Byron, travelling this way in the *Dolphin* in 1764, had been received with exceptional courtesy, although he had described the viceroy of the time 'as absolute a sovereign as any upon earth'. The Honourable John Byron was afforded an impressive welcome with 60 Portuguese officers drawn up in front of the palace and a 15-gun salute. The commodore lived ashore for more than a month. What could not be known was that Louis Antoine de Bougainville on his way to the Pacific the previous year had received a poor welcome when he called in at Rio on his ship *La Boudeuse* and was highly critical of how the Count da Cunha had interpreted the laws of nations in Brazil. This would result in official complaints to Portugal from the French government.

The viceroy, Don Antonio Rolim de Moura, Count of Azambuja, ordered Zachary Hicks to be held in custody. After several hours, when Cook was about to go ashore and demand to know the whereabouts of his officer, a Portuguese boat with a number of officers on board approached the ship. They explained that Hicks would be detained until the vessel's log had been examined, which was a normal arrival procedure. The *Endeavour* was then manoeuvred to an anchorage near the northern end of the Island of Cobras among 30 other ships while Cook waited for permission to proceed ashore to take on water and purchase provisions. A boat filled with Portuguese soldiers now began to circle the ship, keeping her under close surveillance. The gentlemen were anxious to go ashore for the duration of the stay in Rio and collect plant specimens, but word came back that only the captain and his supply boat's crew would be allowed ashore; the passengers would have to remain on board. The seriousness of these orders was emphasized when Banks and Solander donned town clothes and attempted to get away on the pretext of visiting the viceroy. They were turned back by the guard boat and no amount of remonstration could get them past.

This decision was taken by the viceroy because Hicks had been under orders to reveal nothing about the *Endeavour*'s destination to the Portuguese authorities. As she was a ship of war and not of trade, Cook considered that any such enquiries would be impertinent. His lieutenant had carried out the instructions and consequently raised the suspicions of the viceroy about this curious vessel, which, it was claimed, belonged to the British navy but looked like a pirate ship. The authorities would allow only necessary supplies for the *Endeavour*, and the passengers must remain on board. Cook made the suggestion that his ship needed repairs, requiring the gentlemen to live ashore while they were being carried out. The answer was that they could do so, but under house arrest. There seemed to be no way out of this particular bind, except to persuade the viceroy to change his mind by diplomatic notes or 'memorials'. There began a war of words between the captain, only six months into his command, a man of action unused to using the pen for such purposes, and the intractable colonial bureaucracy which knew it had nothing to lose by refusing assistance. Frustration mounted when the *Endeavour* needed to be heeled down for cleaning and caulking her sides and the angle of the deck became so acute that it was almost impossible to walk about. The heat became increasingly unbearable on board. 'Tantalus could never have been more tantalised than we have been at Rio de Janeiro', was how Solander described their predicament. This was a reference to the son of Zeus and the nymph Pluto, who revealed secrets of the Greek gods and was condemned to stand, without food or drink, in water up to his chin under a heavily-laden fruit tree. The mythological association also appealed to Banks: 'Never before had I an adequate idea of Tantalus' punishment but have suffered it with all possible aggravations'.

The Portuguese authorities were deadly serious in their attempts to control the *Endeavour*'s activities, as Zachary Hicks discovered when he was attempting to present yet another memorial to the viceroy. A guard was placed on the pinnace when it reached the shore and Hicks objected, insisting it be removed at once. The result was detention for the seamen overnight in a filthy prison and the officer returned to his ship forthwith. However, surgeon Monkhouse was allowed to go into town to purchase supplies on the daily watering boat, which set Banks and Solander thinking how they could outwit the Portuguese. It was a dangerous game to contemplate because, if they were caught, there was the likelihood of being branded as spies with the severest consequences. Solander decided to take the risk, donned the dress of the surgeon's mate and travelled ashore on the pretence of buying drugs from the apothecaries in town. He was able to acquire some seeds and plants on this rather furtive trip, a rudimentary start to the study of Brazilian flora. From his brief observations Solander made the optimistic estimate that Rio had a population of about 37,000 whites and upward of half a million blacks. He noted that the churches were rich, there were many fine convents, but the opera house's grandeur was apparently not matched by the quality of its performances. Everyone, he considered, black or white, was a slave to the viceroy's wishes. Solander met two Swedish officers who were working in the Portuguese service, but he was unable to speak to them in their common language because they were forbidden to associate with foreigners for fear of revealing information about the fortifications and plans of the city.

Solander's lone trip was, however, an isolated incident, and he and Banks were soon reduced to bribing shore parties to bring back whatever plants they could lay their hands on. The botanists also sorted through the greens that were brought on board as salad for the men and fodder for the animals: 'botanising in company of our sheep and goats', was how Solander described it. A number of interesting specimens were collected in this way and Parkinson and Buchan were able to be kept busy, but it was not sufficient to satisfy Banks' enquiring mind. Nor was it enough for him to circle the *Endeavour* in his small boat, under the watchful eye of Portuguese guards, in rather pathetic attempts to gather molluscs and insects floating out from the land. It was not surprising that Parkinson was able to make paintings and drawings of 22 Brazilian fish because the bay abounded with many varieties and, according to Banks, 'seldom a day passed in which we had not one or more new species brought to us, indeed the bay is the most convenient place for fishing I have ever seen'. He thought, as his line dangled over the side, that this country would be capable of luxuries, were it in the hands of Englishmen, 'as things are tolerably plentiful even under the direction of the Portuguese, who I take to be without exception the laziest as well as the most ignorant race in the whole world'. They had not

The governor's palace at Rio de Janeiro, constructed in 1743.

come all this way just to fish, and finally the frustration of gazing across to shores covered with palm trees and shrubs in glorious bloom got the better of Banks. Under cover of the darkness before dawn, he, Parkinson and a small party lowered themselves from the stern windows of the great cabin into a boat and drifted away from the *Endeavour* on the tide until well out of earshot of the guards.

A whole day was spent on the outskirts of the city, meeting friendly farmers in their small market gardens, where they cultivated cabbages, peas, radishes, pumpkins, water melons and pineapples. Banks and his party were impressed with the profusion of brilliantly-coloured climbers such as *Bougainvillea spectabilis*, with flowers that came in purple, white, red or lilac; there was the rose cactus, *Pereksia grandifolia*, with its clusters of pink or white flowers, and delicate orchids found in shady spots, ready for Parkinson's sketches. Banks was overjoyed to be ashore and he took interest in everything he saw, particularly a hedgerow with many plants in bloom, all of them new to him. There were brilliant poincianas flowering everywhere, and a profusion of colourful butterflies and birds. Banks was able to shoot a loxia but discovered that the little humming birds were impossible to bag. He found the cultivation rudimentary, with bony cattle trying to survive on a type of creeping grass, but the abundance of plant and animal life was almost overwhelming to the visitors after the frustrations of being confined to the ship. Banks tasted the pineapples and gave his expert opinion that they were 'inferior to those I have tasted in Europe; hardly one I have had could have been reckoned among the middling sort, many were worse than I have seen sent from table in England where nobody would eat them.'

The party worked until darkness fell and then returned to the ship. The locals had been aimiable enough, but they talked about their unexpected visitors and word soon got back to the authorities. An armed contingent was sent in pursuit, but they were safely back on the *Endeavour*. However, Banks felt that he and Solander could not risk repeating the escapade. One sorry result of duping the viceroy was the arrest of an Englishman, Forster, who had been acting as liaison between the Portuguese and the British, on a trumped up charge of smuggling goods ashore from the *Endeavour*. He went to prison, his career in Rio de Janeiro finished.

The gentlemen's frustration reached a new pitch when a Spanish man-o'-war entered the harbour freely, without the slightest hindrance, and permission was granted for her officers to live ashore. Cook invited several of them to dine on the *Endeavour* the day after they arrived and in return for the hospitality the Spaniards invited the gentlemen and senior officers to join them at the opera, where they had reserved a box. It was highly embarrassing to explain that the English were confined to their ship.

The tensions began to multiply and manifest themselves in insubordination and lapses of discipline. John Thurman, who had been impressed at Madeira, refused to assist the sailmaker and scored a dozen lashes for his intransigence. Then a seaman, Robert Anderson, attempted to desert and a marine, William Judge, used abusive language to the officer on the watch. Tempers were getting frayed and they would become worse if Cook did not enforce discipline by making an example of miscreants. The boatswain's mate, John Reading, went about his flogging of Anderson and Judge in a desultory manner and, in a chain reaction of punishment, he too found himself on the receiving end of a lash.

Banks was of course unable to make full notes about the city, but James Mario Matra, because of his many visits to the town on the provision boat, left a good description of Rio, and it would be hard to guess from his writing that there was any embargo on the *Endeavour*'s men going ashore. He observed that the life of the city revolved around the governor's palace on the waterfront, which was part of the royal square, also containing the palace, mint, stables and jail in an impressive two-storey building. The left wing of the square included shops, which faced the fountain supplying water to both visiting ships and the local residents, the area being constantly crowded with blacks waiting to fill their jars. The market place extended from the north-east corner of the square along the shore, being very convenient for fishing boats and those carrying produce from the other side of the harbour. The vendors were all blacks who in quiet moments sat spinning cotton. The gentry of the town travelled in carriages drawn by mules and the ladies had sedan chairs, boarded fore and aft, with curtains on either side, carried by two slaves. The apothecaries' shops served as meeting places for the gentry and the coffee houses provided places of relaxation with backgammon a favourite pastime. In spite of the heat, the gentlemen dressed stylishly for their town appearances, but tended to wear less when in the privacy of their homes. Matra noticed an absence of beggars in the streets, unlike Europe, and prostitution seemed to be carried out through assignations in the many churches scattered around the central area. It was said to be quite common for husbands to send their wives to mass at 2 or 3 a.m. and then use the period of their absence for dalliance and intrigue 'which those early hours afford'. The town contained several courts of justice presided over by the viceroy, who had a council appointed from Europe to assist him. Matra noted also the whale fishery that supplied Rio with lamp oil, the imports of brandy from the Azores, wine from Madeira, European goods from Portugal and slaves from Africa.

The *Endeavour* was made ready to sail by 2 December and Banks wrote in his journal, 'This morning, thank God, we have got all we want from these illiterate, impolite gentry, so we got up anchor and sailed to the point of the Island of Cobras, where we lay and wait for a fair wind which should come every night from the land'. But it was not forthcoming that night or the next, and they remained in the harbour. Their manoeuvres were marred by the death of able seaman, Peter Flowers, who fell from the rigging and drowned before he could be reached. 'A good handy seaman who had sailed with me for five years', Cook wrote. He was replaced for the journey ahead by a Portuguese sailor named Pereira. Another multiple flogging was carried out when it was discovered that the gunner, Stephen Forwood, in charge of a watch with two others had broached the spirit cask on the quarter deck and taken about ten gallons of rum, some of which was found in his cabin. The three men were punished with twelve lashes each and Cook stopped their allowance until the stolen amount was made up. The arrival of a Spanish brig and the offer from its commander to carry letters to Europe set Cook, Banks and Solander writing to the Admiralty and the Royal Society giving rather wounded accounts of their unreasoanble treatment at the hands of the viceroy. 'The manner we have been received and treated here is such as was never practiced on any English ship', fumed Cook. Linnaeus in faraway Sweden was informed of their plight by Solander: 'We have had to argue and bully our way for everything that we have got here. The viceroy is a rather obstinate and stupid man'. No personal notes were sent from Rio by Banks, and Harriet Blosset would have to spend many more months at her embroidery before she learned of her intended's progress on his grand tour.

The recorded history of Brazil's flora went back to the sixteenth century with an account by J. Lorg published in *Niederlassung der Franzosen in Brasilien*. Later accounts came from George Marcgrave, who was in Brazil from 1638 as a member of a scientific team based in Recife. The Englishman, William Dampier had also collected there as part of his voyage around the world at the end of the seventeenth century and he took back plants to the Sloane Herbarium and to Oxford University. By the time of the *Endeavour*'s visit, Banks had overlooked Dampier's collecting trip, or chose to ignore it, when he wrote in his journal, 'No one that I know of even tolerably curious has been here since Marcgrave and Piso about the year 1640, so it is very easy to guess the state in which the natural history of such a country must be'. In spite of the very limited collecting opportunities, Banks' plant specimens from Brazil totalled 320 species of which about 112 were new to science. Sydney Parkinson was set to work illustrating 37 of them. Twenty-three plates engraved from these watercolours would survive for inclusion in the *Florilegium*.

The breezes remained unfavourable on the broad expanse of the harbour and Cook decided to launch two of his boats in an attempt to tow the *Endeavour* out to sea. As they slowly passed the fortifications at Santa Cruz near the entrance to the bay, two shots were fired at them, one almost hitting the mainmast, and Cook had to go across to the battery to find out

Aechmea nudicaulis Watercolour drawing of a plant which is pollinated by humming birds, collected at Rio de Janeiro.

what was happening. The officer in charge had received no notification that the ship could depart and a further delay was experienced while inquiries were made. Finally, they were allowed to leave with the viceroy's rather hollow written wishes of *bon voyage*. It had taken five days to get out into the ocean but Cook had used the time to good advantage, making an extensive chart of the harbour, noting all the fortifications, while Alexander Buchan produced profiles for this comprehensive survey. And then, on 7 December, the captain was able to record in his journal, with admirable restraint, 'Now we are fairly at sea and have entirely got rid of these troublesome people'.

3
Rio de Janeiro — Tierra del Fuego
1768–1769

They change their clime, not their frame of mind,
who rush across the sea.

Horace (65–8 BC)

CHRISTMAS Day 1768 saw 'Mrs Endeavour', as Banks had affectionately dubbed her, proceeding down the coast of South America with steady progress south by west in the broad Atlantic swells. Solander, Parkinson and Spöring had been kept busy classifying, noting and drawing the plants from Rio de Janeiro. Banks spent some of his time shooting birds and collecting marine life. On this day, however, it was a time for leisure and reflection as the table in the great cabin was cleared of the study materials and a celebratory dinner was held followed in the evening by fine madeira wine, port, brandy and cigars. The officers and gentlemen sat and reminisced about Christmas past and drank toasts to their loved ones at home. It was exactly four months since they had set out from Plymouth and, apart from the troublesome business far behind them in Brazil, the voyage had been satisfying and successful.

The festivities were less restrained and nostalgic on the other side of the bulkhead forward. A continuous, noisy party was under way down there with copious supplies of rum to bolster the seasonal mood of this special day. It continued at high pitch until the alcohol finally took its toll: 'There must scarcely be a sober man on the ship', Banks observed to Cook, 'thank God the wind is very moderate or the Lord knows what will become of us!' With characteristic understatement the captain admitted, allowing a smile on his normally taciturn features, 'Indeed the crew are none of the soberest'.

By now the air temperatures were falling as they moved further south and, although it was the height of the southern summer, heavier clothes were necessary in these higher latitudes, particularly after spending so long in temperate and tropic conditions. All hands were issued with Magellan jackets, made of a thick woollen fabric, or Fearnoughts, clothing made of padded canvas. Banks put on a flannel coat and waistcoat over thick trousers. The officers, warrant officers, mates and midshipmen received only trousers to help ward off the cold and damp; they were expected to

34

provide all the other protective clothing themselves. For several evenings swarms of butterflies, moths and other insects flew about the rigging, blown from the shore about 30 leagues away. Thousands of them settled on the vessel and Banks asked the men to gather them up for him. After selecting the specimens he found interesting, the rest were thrown overboard and his helpers rewarded with bottles of rum. With fresh gales and clear weather at the end of the year, whales and porpoises played around the ship and there were great shoals of shrimp colouring the seas all around with a ruddy glow. On New Year's Eve, a relaxed captain and his officers joined the gentlemen around the great cabin's table to toast the year ahead. 'Success to sixty-nine!' The words were repeated and they drank reflectively under the gently swinging oil lamps, realizing that, God willing, it would be as exciting a year as any had experienced before. Outside, large clumps of weed floated on the surface of the sea, which in daylight now displayed the aquamarine tints of southern latitudes.

It was necessary to make a landing somewhere in these parts to take on firewood and water before tackling Cape Horn. Sydney Parkinson said he would welcome a call at Port Egmont, the British settlement in the Falkland Islands, to send home some mail, and Banks was anxious to add that territory to his own botanical map of the world. Cook decided, however, with austere authority, that he should not chance delays by such a diversion and stayed well to the west. It was a great disappointment to Joseph Banks and he complained that he would 'not now have a single opportunity of observing the produce of this part of the world', and it became the first of several such disagreements between the captain and his paying passenger. Cook faced the considerable challenge of negotiating the formidable obstacle of the Horn and arriving at his Pacific destination in time for the transit of Venus, which was the principal reason for this voyage, and he considered the dilettante devices of botanists must be afforded a lower priority. In fact little was lost by not visiting the Falklands, which had been described by Bougainville, on his visit there four years earlier as being 'a vast silence and a sad and melancholy uniformity'. Any petulance about missing this destination soon subsided in the discomfort of a hard gale, which blew so heavily that the *Endeavour* had to be brought to under a mainsail. The violent motion dislodged Banks' wooden bureau, causing most of the books it contained to be strewn about the cabin floor. The roar of the wind and waves, the shuffling of the books and rattling of cots against the sides of cabins made for a sleepless night. At least the senior officers and gentlemen had the privacy of two small lavatory buckets just outside the great cabin, which could be used without having to brave the elements. For the rest of the crew there were only the heads: platforms suspended over the ship's side, requiring certain basic skills when urinating or defecating into the wind.

After the Falkland Islands disagreement, Banks was excited to see the coast of Tierra del Fuego one morning in early January, and relieved to find that it looked by no means the barren shore he had been expecting from the reports made on other voyages. Cook agreed with Banks that he could land if a suitable place were found. An opening was discovered and the *Endeavour* was steered to within a mile before being forced to stand off again as there was no apparent shelter. The botanists eagerly scanned the country through their telescopes and noted spots of white and yellow which they judged to be flowers; the white in large clusters and the yellow in dispersed areas on the side of a green hill. Banks wanted to disembark at what was named Vincent's Bay but there were rocky ledges in the way covered with seaweed and a difficult bottom for anchoring with depths ranging between four and nine fathoms. Cook decided to find a more convenient and safe harbour for taking on his wood and water in the strait between Staten Island to the east and Tierra del Fuego to the west, but he was persuaded to wait while the botanists went ashore. He noted, 'At nine they returned on board bringing with them several plants and flowers most of them unknown in Europe and in that alone consisted their whole value'. This waspish journal entry about their activities was all the captain allowed himself to reveal of his real feelings; his other thoughts were kept as contained as a closed book. Banks and Solander had to negotiate heavy surf to get ashore but the effort was worthwhile because they gathered a hundred new species, including white and red cranberries, winter's bark with medicinal properties and sea celery, which the cook would prepare as fresh greens.

Shortly after this excursion the *Endeavour* was taken into the wide Bay of Good Success, accompanied by frequent showers of snow and rain. Thirty or forty natives appeared on the shore but ran away when a party from the vessel approached them. Banks and Solander went ahead on their own and two of them approached with caution, threw away the sticks they had been brandishing in an apparent indication of peace, and met their visitors on the beach. Others hovered in the background as three were invited to the ship. Once on board they were given jackets to cover their almost naked bodies. They were also offered some bread and beef to eat but seemed unimpressed with the food, thrusting the uneaten portions under the new garments. Rum and brandy were proffered but it burned their throats and they choked, prompting Parkinson to view the reaction as confirmation of his teetotal opinion that 'water is the most natural, and best drink for mankind, as well as for other animals'. The natives then made it known by sign language that they wanted to go back to the shore and rejoin their fellows after all attempts at communication had failed; not a word had been understood by either side. These Fuegans were very light fingered and one of them stole the leather cover of a globe in the great cabin, hid it under his skins and carried it ashore undetected. Once there he revealed the prize, waving it with pride of possession and placing it on his head like an absurd tea cosy. This was to the huge amusement of the crew and the annoyance of Banks, whose efforts to retrieve it were without success.

It was the first encounter with primitive people for most of the travellers and set Alexander Buchan to work sketching figures and the rather bleak environs of the bay where they lived while Parkinson, Banks and Cook made extensive entries in their journals, noting that the local natives were nomadic hunters, looking uncouth and savage with broad, flat faces, small eyes and low foreheads. Their straight black hair hung down over the ears and was smeared with a red colouring. The men were quite tall, averaging 5 feet 10 inches and wore the skin of seals or the guanico — a South American llama — wrapped around their shoulders, leaving one arm exposed. Both men and women displayed necklaces made from little periwinkle shells plaited in rows with dried grass. The women, who were much shorter than their men, wore a flap of skin around the thighs and carried the children on their backs. Parkinson wrote that they were generally engaged in domestic drudgery at the small village on the south side of the bay, where

Fucus giganteus.

Sydney Parkinson pinx '769.

they were living with their dogs in a cluster of beehive-shaped huts. These were made of tree branches covered with the same guanico and seal skins used for the skimpy clothing, and regarded by the Europeans as wretched, smelly habitations. The natives' main means of sustenance seemed to be seafood: seals, shellfish and mussels caught from the shore, because there were no boats in the vicinity for fishing. Cook observed some of the natives with rings, buttons, cloth and canvas of European manufacture, which suggested to him they must travel to the north, 'as we know of no ship that has been in these parts for many years'. They were obviously familiar with firearms and signalled to the crew to fire at seals and birds. Cook thought them 'perhaps as miserable a set of people as are this day upon earth',

Macrocystus pyriferus Watercolour drawing from Tierra del Fuego by Sydney Parkinson.

without boats, few personal possessions, armed with bows and arrows and having no apparent hierarchy.

On inspecting the high hills surrounding the bay through a telescope, it looked as if they were capped with grass above thickly-wooded lower slopes. Banks was certain there would be some interesting alpine flora up there and he organized an expedition to find out. Charles Green and William Monkhouse agreed to join his party of 12, which also included servants and seamen to help carry the guns, collecting bags and refreshments for a day's excursion. They set out just before dawn to take advantage of every moment of daylight and entered the lower woods as the sun was rising above the horizon in a great yellow blaze. 'Just like a sunshiny May day in Lincolnshire', Banks remarked to Solander as they trudged upwards, having to force a path through dense thickets where it was likely no humans had ever been before. The overweight Solander grunted in agreement and climbed on, panting. Stops were made on the way to collect a few plants, eat an early cold lunch and shoot a vulture that suddenly appeared overhead. But the ascent was taking much longer than anticipated because, unknown to them, they were travelling in a huge semi-circle rather than by the most direct route, their vision obstructed by dense greenery. By the time the party was clear of the trees it was already 3 p.m. and what had appeared from far below as grass, turned out to be birch scrub growing on very boggy ground. Banks decided to push through, although it was hard going having to negotiate the waist-high shrubs with feet sinking ankledeep in the spongy surface. A mile later they reached firm ground, still in reasonable spirits, but increasingly tired from the taxing climb. The return journey, however, would be downhill and because it stayed light until quite late there was little problem with time. Suddenly, without warning, Alexander Buchan collapsed in a violent fit, his first since leaving England. It passed quickly but left him very weak. Banks ordered a fire to be lit so that the shivering artist could rest and recover. The sailors and servants stayed with him while Banks, Solander, Green and Monkhouse went ahead to collect the alpine plants, the main objective of the expedition.

The late afternoon weather began to turn chilly with icy winds and snow-showers replacing the bright sunshine. The light faded earlier than anticipated and Banks' infectious enthusiasm turned to disquiet as he realized it would now be impossible to reach the ship before darkness. The only solution to this predicament was to head for the woods and build a fire where they could all shelter until daybreak. Buchan's state had improved, much to Banks' relief, although like everyone else he was feeling the effects of the penetrating cold. The birch bog was re-negotiated and Banks stayed at the rear of the group as they stumbled on, urging them to maintain a lively pace. There was a sense of apprehension in the air, but no real danger, until the stocky Solander slipped and sank to the ground in exhaustion. 'I am sorry', he muttered, 'I cannot go any further, I have to lie down'. Banks barked at him, 'You must not do that. Look! the ground is getting covered with snow. You will freeze.' Solander seemed to be deaf to his warning and remained prostrate where he had fallen. When the black servant Thomas Richmond began to show similar symptoms, Banks sent Buchan ahead with half of the party to prepare a fire in the woods while he arranged to have Solander and Richmond carried down to its warmth. There was a delay of fifteen minutes, during which Solander slept soundly and Richmond's condition worsened, but in that time the advance group had succeeded in coaxing a fire from damp wood that littered the ground and constructing a rough shelter from tree branches. A seaman and the other black servant, George Dorlton, were told to stay with Richmond, who could not be moved, while Banks, combining pushing and carrying, bundled Solander down the quarter-mile track to the fire. Afterwards, two of the men went back for Richmond but in the failing light and thickening snow they could find nobody in spite of combing a wide area and calling out their names.

While they were away Banks discovered that a bottle of rum was missing from the provisions bag and concluded, correctly, that the sailor and servants must have drunk it and then fallen asleep. The blizzard worsened and all hope had been lost for the missing members when, at about midnight, shouts were heard coming from the nearby trees. It was the missing sailor, who had been aroused from his alcoholic stupor by the intense cold and stumbled down the hill in the darkness, attracted like a moth to the flickering flames. Banks ordered men to return up the hill with the sailor to locate the servants, and they did find them, but both Richmond and Dorlton were unable to walk and had to be left where they were, protected a little from the elements by a shelter of leaves and branches. It was a long, sleepless night around the smoky fire waiting for the first light of dawn. Their situation was grave: two of the party were probably near to death, others were faltering and there was always the possibility that Buchan's epilepsy would recur. At daybreak they set out again, through snow that was already beginning to melt and found the two servants frozen to death with Banks' greyhound bitch standing guard over her lifeless keepers.

After making a meal of the vulture they had shot the previous day, the sadly disillusioned party made its way back to the ship. They were on board by late morning after a descent of only three hours, much to the relief of Cook and his officers.

Parkinson was to describe the plants and curiosities collected at the Bay of Good Success as nondescript, although the others were most impressed with the flora and Banks claimed that it set him wondering 'at the infinite variety of Creation'. It was a theme eagerly embraced by Parkinson who thought that no humans could survey the majesty of the bleak landscape around them with its snowcapped peaks without shuddering and considering how amazingly diversified were the works of God 'within the narrow limits of the globe'. Cook was in no mind for any sort of philosophy; he was being characteristically practical, surveying the bay while his men collected wood, water and the edible wild celery that would be mixed with the crew's ration of pease or added to ground wheat and portable soup. Banks and Solander attempted to classify a considerable variety of trees, including the beautiful Antarctic beech, *Nothofagus antarctica*, which Solander thought a type of birch, various shrubs such as *Berberis ilicifolia* with orange-yellow flowers and holly-like leaves, and a range of perennial herbs including the wild celery, *Apium prostratum*, and the stubby green *Plantago barbata*, a native of this area. The *Endeavour*'s guns were stowed below decks in anticipation of the stormy waters of the Horn and the ship was made ready for sea once again. They weighed anchor on 21 January, making their way westward out of the Bay of Good Success and along the Strait of Le Maire. Banks' main concern was that the seas should remain moderate for long enough so that classification could be completed on the new plants, which were stored in keeping boxes. Although Parkinson, Spöring and Buchan were now so used to the motion that Banks knew 'it must blow a

Beehive-shaped dwellings at Tierra del Fuego. Engraving after Parkinson.

gale of wind before they leave off', they were in fact approaching waters notorious for their gales. Banks was excited about the collection because it came from areas where few had botanized before. The plants of the Falkland Islands and the Straits of Magellan had been studied superficially during Bougainville's voyages in 1763 and 1764, and the Frenchman had visited Tierra del Fuego again on his way to the Pacific in 1767 with the naturalist Philibert Commerson on board. The fact of being among the first Europeans to examine this flora was exhilarating for Banks and Solander. They had collected 148 species at this southern tip of South America. Parkinson illustrated 70 of them, of which 65 eventuated as engravings for the *Florilegium*. The memory of the recent expedition, with its loss of the two servants, was pushed to the furthest recesses of Banks' mind and he left this continent thinking that 'probably no botanist has ever enjoyed more pleasure in the contemplation of his favourite pursuit than Dr Solander and myself among these plants'.

Doubling the Horn, as sailors called it, was a monumental challenge for any navigator wanting to enter the Pacific from the east. There were several ways of doing it and each was fraught with peril. The most sheltered route seemed to be the Strait of Magellan to the north of Tierra del Fuego, but there was scarce sea room for a lumbering square-rigger attempting to beat

40

westward for most of the way through twisting waterways, frustrated by irregular downdraughts from the mountains on either side. If that was not enough, there were no reliable charts of the passage and few safe anchorages, as Cook's predecessors had discovered. There was an unnamed channel even narrower than the Strait of Magellan, between the southern coast of Tierra del Fuego and the mass of islands spreading to the south and terminating in Cape Horn, but it appeared on no charts and there was little time for Cook to be exploring the maze of inlets and islands of this inhospitable place. The captain could only form an opinion of the best way to reach the Pacific by relying on his own intuition and reading the recent journals of Byron and Wallis. Commodore Byron on the *Dolphin* had spent more than seven weeks passing through the Strait of Le Maire in early 1765 and then Captain Samuel Wallis, commanding the same vessel on her next voyage, was forced to remain in the area for three months attempting to reach the Pacific against contrary winds. From an account of George Anson's voyage around the world on the *Centurion* published in 1748 Cook had taken particular note of a chapter headed 'Observations and directions for facilitating the passage of our future Cruisers round Cape Horn'. Anson encountered terrible March storms and recommended that by standing far south of the land, difficult currents would be avoided and the weather might be less stormy and uncertain. Cook chose to follow this advice and his track would become a masterpiece of navigation: through the Strait of Le Maire to the east of Tierra del Fuego and then south of the Horn. This meant more than 1,000 extra miles of westing before the South American continent could be cleared and the possibility of westerly gales, but there was plenty of room to manoeuvre down there with no land to run on to. Once past the almost symbolic Horn it should then be plain sailing to the north and west deep into the Pacific. Summer offered no guarantee of an easier passage in these waters, but James Cook made steady progress to where his and Charles Green's calculations indicated the notorious Horn should be. The weather was foggy and from three leagues off they passed what looked like an island on the starboard bow with a high rough hummock on it. There was no other land to the south or the west and the captain, the astronomer and senior officers agreed they had doubled the Horn with as little danger as rounding the North Foreland on the coast of Kent.

Periods of fine sailing alternated with calms so that Banks could get out in his little boat and shoot seabirds for his growing collection. He recalled the words of the Athenian legislator, Solon, who wrote: 'By wind the sea is lashed by storm, but if it be unvexed, it is of all things most amenable'. Albatrosses were everywhere and, as there was no superstition among the naturalists about killing them, several fell foul of his musket and provided what the gentlemen, at least, regarded as a thoroughly acceptable meal, preferable to salt pork and memorable enough for the recipe to be recorded in Banks' journal: 'Skin them and then overnight soak their carcasses in salt water till morning, then parboil them and throw away the water, then stew them with very little water and when sufficiently tender, serve with a savoury sauce'. Eccentric dishes such as stewed albatross only emphasized that the type of food served on British ships of war during long voyages had changed very little since the times of Prince Henry the Navigator, the leading promoter of exploration during the Middle Ages. This was due to sailors' trenchant resistance to change, the chronic problem of corrupt suppliers, and the fact that preservation of food relied almost exclusively on salt. The British were less ingenious than any other nation in trying to introduce a more interesting diet at sea and improvements in navigation far outstripped any developments in diet. The Dutch took many varieties of their cheeses on long voyages, together with pickled and smoked herrings, the Italians had olive oil and pasta in all shapes and sizes, the French were able to bake excellent fresh bread at sea, the Spanish took sardines, garlic and dried tomatoes with them and the Portuguese carried olives and dried codfish. Samuel Pepys, the great naval administrator and diarist observed that 'Englishmen, and more especially seamen, love their bellies above everything else' and yet they made do with generally atrocious fare. A pamphlet published in London by William Thompson in 1761 described the food on British warships as including bread teeming with large black-headed maggots so that the men had to shut their eyes before they could bring themselves to swallow it, stinking beer as abominally foul as the human excrement pumped from London cellars at midnight, and salt beef discovered to be already putrid when taken from sealed barrels. Sometimes the preserved meat looked suspiciously like horse flesh and sailors claimed that horseshoes often lurked in the bottom of casks.

A typical ration for a crewman was a pound of biscuit a day, two pounds of lean salt beef twice a week, a pound of salt pork twice a week, half a pound of pease four days a week and smaller amounts of dried fish, butter and cheese. The water on board quickly became unfit to drink because it was impossible to keep fresh in wooden casks and a gallon of beer was allowed each day in its place, while it lasted. This diet provided enough energy for a hard-working seaman but it was likely to bring on the scurvy very rapidly. There had been sufficient evidence from individual naval commanders for centuries that fresh fruit, and particularly citrus, was effective in combating the terrible affliction. However, it was not until the mid-eighteenth century that the Scots naval surgeon, James Lind, began a serious study of the problem. He was spurred on by the completion of Anson's circumnavigation, during which more than half his crew died from scurvy. Lind, who would become known as 'the father of naval medicine', began a series of controlled experiments with various groups and the results were published in his treatise of 1753, suggesting that immediate relief was possible by taking orange and lemon juice and, to lesser effect, cider. Lind's clinical trials were a model of efficiency and set procedures for many future medical studies, but the Admiralty was not interested in the findings and the Navy's Sick and Hurt Board rejected his conclusions outright. 'The promise has been mine to deliver', was Lind's resigned comment, 'the power is in others to execute'.

James Cook inherited the woolly thinking of the Lords of the Admiralty and, in spite of Lind's evidence, of which he knew nothing, was instructed to make his own experiments with a number of possible remedies, including malt. In July 1768 the Admiralty had written to him requesting trials to be made according to the writings of Dr David McBride as set out in his pamphlet *An Historical Account of the New Method of treating the Scurvy at Sea*. This concentrated on the consumption of malt and involved a rather complicated procedure of converting it into a liquid and then making a panada, a sort of bread pudding, using sea biscuit and dried fruits. Scurvy sufferers were to take two meals a day of the panada together with drinking liquid malt. The *Endeavour*'s surgeon, William Monkhouse, was charged with keeping an exact record of the results of treating scorbutic and other diseases and instructed to send his journal to the Admiralty at the conclusion of the voyage. Banks had sought his own advice on the subject back in London and carried with him a special order of concentrated orange and lemon juice as medicine. The lemon was evaporated by a third in a two gallon cask, another contained a mixture of seven gallons of orange juice mixed with one gallon of brandy and a third container held five quarts of lemon juice mixed with a quart of brandy. The Hatton Gardens suppliers made sure that their client would enjoy indulging in his scurvy remedy.

It was fortunate that Cook had what seemed to sailors an abnormal preoccupation with cleanliness, fresh food and green vegetables as being the way to ensure a healthy ship. It was an instinctive feeling that proved correct, since the *Endeavour* remained remarkably free from the dreaded scurvy. One of the problems of supplying citrus juice was its cost, because supplies had to be brought all the way from the West Indies, and the Admiralty was notoriously pennypinching when it came to good food and expensive provisions. This attitude extended even to the calibre of cooks, who were regarded as only minimally important. Standing instructions for sea-cooks were brief, and to the point: to look after the meat in the steeping tub, to prevent it being lost in rough weather, to boil the provisions and deliver them to the men. That was the extent of the stated duties, which meant that cooks were not expected to have any qualifications, in fact, the fewer they had, the more likelihood of getting the job. The task on the *Endeavour* was to feed more than 90 men every day. The first cook appointed at Deptford when the ship was being commissioned proved to be lame and infirm and Cook would not accept him. He had then chosen his own suitable replacement but the Admiralty insisted that their appointment, John Thompson, must take the position on the *Endeavour*. As it turned out, Thompson with his one arm proved to be thoroughly acceptable and even Joseph Banks, who was no stranger to the best cuisine London could offer, appreciated his culinary ingenuity, particulary his finesse with seafood and an impressive variety of birds. Cook's instructions to experiment with anti-scorbutic foods and substances put an extra load on Thompson, who had to handle sauerkraut by the barrelful, hogsheads of malt and the portable soup, which was meat extract rather like large stock cubes. But he was always cheerful. Cook was convinced that regular servings of malt, sauerkraut, and portable soup kept the scurvy at bay and 'by this means and the care and vigilance of Mr Monkhouse the surgeon this disease was prevented from getting a foothold on the ship'. The sharp-tasting sauerkraut was at first rejected by the men. The captain, rather than forcing its consumption by corporal punishment, employed more subtle techniques to persuade them. He ordered sauerkraut to be served each day to his officers and left it up to the men either 'to taste as much as they pleased or none at all'. It took only a week to have the entire complement eagerly taking the previously unpalatable soused cabbage, 'for such are the tempers and dispositions of seamen in general'. Cook had won his way by simple psychology, making sauerkraut desirable by example. As he stated, with an amused smugness over dinner in the great cabin, where the rather unpalatable stuff was in reality not so highly prized, 'The moment they see their superiors set a value on it, it becomes the finest stuff in the world and the inventor a damned honest fellow'.

At the end of January they reached 60 degrees south, their highest

The *Endeavour* at sea by Jack Earl, 1970.

southern latitude and surprisingly the weather was not cold and gloomy as might have been expected, but pleasantly bright during the long hours of daylight. Cook now altered his course, steering NNW into an alternation of gales and fair weather with the happy prospect of warmer days ahead. During the third week of March, with the sea becoming bluer and the air growing, as Sydney Parkinson sensed, 'dry, serene and salubrious', numbers of tropical birds visited. Two were shot and recovered from the sea for study. By now the men were used to the regular discharge of Banks' firearms and they rarely looked up from their crowded working spaces on the main deck to see the results of his fusilades. Regular practice was also held with the ship's guns and firearms while the marines maintained their military disciplines of parades and practice together with the 24-hour guard they mounted on key areas of the ship, such as outside the powder magazine or the captain's quarters.

Cook ran a very tight ship on this, the longest ocean passage he had undertaken, and most insubordination was kept in check by the constant threat of floggings. Two months after leaving Tierra del Fuego, however, something happened to demonstrate how human emotions could erupt unexpectedly in such confined conditions. A young marine named William Greenslade was on guard duty outside the great cabin when he was asked to look after a piece of sealskin that belonged to one of Cook's servants,

William Howson, who was suddenly summoned away. Greenslade had been wanting to buy some of the skin from him to make a tobacco pouch, but Howson had repeatedly refused. Now, without thinking twice, Greenslade cut off a piece and pocketed it and when Howson came to retrieve the skin he was furious at this violation of his property. He complained to the other marines, who began to taunt their colleague about how his action besmirched the honour of the corps. John Edgecumbe, their sergeant, threatened to inform the captain of the incident and Greenslade knew that he was for the lash. The sensitive young man agonized about his fate for the rest of the day and when he was called to the deck at dusk by the sergeant, he went forward on the opposite side of the ship and was assumed to have gone to the heads before giving himself up. Half an hour passed before they realized that William Greenslade had thrown himself overboard into the watery void; it was far too late for any hope of rescue. This black incident left the marines one man short and impossible to replace. They were approaching unknown regions where their full support might be needed to guard the *Endeavour* from threats by hostile natives and the officers and gentlemen against any mutinous actions from within. For a moment, the Age of Enlightenment had reverted to aberrant barbarism.

European contact with the Pacific began about 250 years before the *Endeavour* entered the world's largest ocean. In the sixteenth century Spain's empire had spread across Europe and the Americas with its fleets ruling the seas. Vasco Nuñez de Balboa was the first European to view the great ocean when he travelled overland across the Isthmus of Panama and on 25 September 1513 sighted *Mar del Zur*, the great South Sea, claiming it for his king in Madrid. The first to travel across the ocean was a Portuguese in the service of Spain, Ferdinand Magellan, seeking a short route to the fabled Spice Islands, the Moluccas, by sailing in a westerly direction instead of east. He embarked in August 1519 with a convoy of five ships travelling far south and eventually found a way into the South Sea through the strait named after him. His story is one of immense danger and hardship; more than 150 of his men died as they crossed the Pacific and Magellan himself was killed by natives in the Philippines, where no European had been before him. His flagship, the *Victoria*, eventually returned to Spain after an absence of more than three years, the first to circumnavigate the globe. The question of a great southern continent, the balancing land mass at the bottom of the world, was thought to be solved when Pedro Fernandez de Quiros, another Portuguese, stated that the Solomon Islands, discovered by Mendaña in 1568 were, in fact, Terra Australis. Quiros sailed from Peru in December 1605 with Luiz Vaez de Torres as the captain of his second ship. While de Torres sailed through the reef-filled strait between Australia and New Guinea which now bears his name, Quiros went on to the New Hebrides, naming them 'Australia of the Holy Spirit'. The discovery of New Guinea as an island was reported, and ignored, while Quiros was determined to explore what he imagined to be the new southern continent. But Spain had little interest in anything but protecting the vast treasure houses she had already discovered and was certainly not in an expansionist mood while English buccaneers, who intercepted Spanish treasure galleons on their regular voyages between Acapulco and Manila, needed to be repulsed.

The *Mar del Zur* had become international, with some highly desirable prizes to be won. English interest was sparked in the late sixteenth century following Francis Drake's circumnavigation of the globe and return to Plymouth laden with booty from Peru, for which he was knighted by Queen Elizabeth. Expanding Dutch interests in the riches of the East Indies saw their ships rounding Cape Horn, which was named by them after the town of Hoorn in Holland. Later, William Dampier was probably the first Englishman to set foot in Australia and his writings about distant discoveries became popular books. Subsequent British circumnavigations by George Anson (1741–1744) and the Hon. John Byron (1764–1766) resulted in a few discoveries. On Byron's return, the Admiralty had ordered Samuel Wallis to cross the Pacific by steering far to the south, in search of Terra Australis Incognita. Instead, he discovered Tahiti, although he claimed to have glimpsed the coast of a great continent to the south. It was now left to James Cook, unaware that Louis Antoine de Bougainville had preceeded him, to try his luck with the additional political purpose of his voyage — that of discovering and annexing territories, perhaps Terra Australis itself, for Britain's expanding empire.

4
Tierra del Fuego — Tahiti
1769

I am as free as nature first made man,
Ere the base laws of servitude began,
When wild in woods the noble savage ran.

John Dryden (1631–1700)
The Conquest of Granada

AT the beginning of April they experienced the welcoming feel of the South Pacific, with blue seas, warm airs and tropical birds all around the ship. Those who had been this way before, and there were a few from Byron's and Wallis' voyages, could sense the proximity of land. On the morning of 4 April, with the *Endeavour* continuing on a steady north-westerly course between the first and second tracks taken by the *Dolphin*, Banks' servant, Peter Briscoe, who was on duty as a member of the second watch, sighted an island on the horizon. They took two hours to reach within a mile of it, with rising excitement from the ship's crew. This was Cook's first Pacific discovery, a crescent-shaped atoll enclosing a broad lagoon with great clumps of palm trees at one end. The details became more apparent as they moved closer; smoke rose in several places and it was interpreted as signals from natives but night fell before the western end of the island could be examined and, not knowing if there was other land in the vicinity, the ship lay to.

The following morning saw outrigger canoes in the lagoons and natives standing on the reefs watching this surprise visitor from another world. Banks and Solander waved for them to come out to the ship, but they would not budge. The President of the Royal Society, Lord Morton, had prepared a series of 'Hints' for the *Endeavour* voyage relating to the treatment of indigenous peoples and the procedures recommended for observing the transit, and Cook studied these carefully as the ship neared its destination. The gentlemen, officers and crew were requested 'to exercise the utmost patience and forebearance with respect to the natives of the several lands where the ship may touch. To check the petulance of the sailors, and restrain the wanton use of fire arms. To have it still in view that shedding the blood of those people is a crime of the highest nature'. He also advised regular prayers. Cook was sympathetic to the President's hints about natives and astronomy but he was not very interested in religious matters and prayers were usually left to his officers.

On 10 April a high island appeared and was identified as Wallis' Osnaburg by a profile that appeared in his journal. The next morning other outlines drawn on the *Dolphin*'s last voyage indicated that the towering peaks on the horizon, enveloped in cloud, were their destination, King George's Island. As they got nearer, a few natives came out to the *Endeavour* seeking to trade coconuts and apples for nails, buttons and beads which they seemed to know from previous European visits were likely to be carried on board. They were not afraid of the visitors in the huge vessel, although declined to go aboard in spite of their friendship salutes of 'taio, taio'.

Wind squalls and heavy rain hampered progress, forcing the vessel to lay to that night. Next day, guided by the pinnace, the *Endeavour* glided through the gap in the reef charted by Wallis and came to anchor in 13 fathoms of clear water half a mile from the shore. The land in front of them seemed to Parkinson as uneven as a piece of crumpled paper, divided into irregular valleys and hills, all covered with a thick blanket of dark green vegetation from the shoreline right up to the highest peaks. They were eight months out of Plymouth and almost two weeks ahead of schedule. Many canoes came to greet this, the latest arrival in what had become a regular sequence of pale-faced visitors: Wallis in 1767, Bougainville in 1768 and now Cook in 1769. They came to trade bananas, coconuts, breadfruit, apples and hogs, although there was much less produce offered than Wallis had reported. Some of the natives came on board and Sydney Parkinson soon discovered what excellent pilferers they were when an earthenware pot from his cabin was missing after a bartering session for some cloth. The Tahitians, as they would be called much later, were generally friendly and men that had been on the *Dolphin* were remembered and greeted warmly. Many natives had assembled on the beach and as the *Endeavour*'s officers and gentlemen stepped ashore with the marines drawn up in marching order every man was presented with a green plantain or banana frond. An area of ground was cleared where the Tahitians placed their leaves and the visitors were directed to follow suit in a ceremony of peace. They walked together for four or five miles under the shady groves of coconut and breadfruit trees. Native huts, mostly without walls, clustered under the trees and Banks was overwhelmed by the serenity and beauty: 'In short the scene we saw was the truest picture of an Arcadia of which we were going to be kings that the imagination can form'.

Cook needed to maintain firm control over his trading goods, keeping enough in reserve for whatever future encounters with primitive peoples might take place in the Pacific. He also had to ensure that relations with these natives did not get out of hand, with the prospect of at least two months' stay. Before arriving at Matavai Bay he had issued his 'Rules to be observed by every person in or belonging to His Majesty's Bark the *Endeavour*, for the better establishing a better and uniform Trade for Provisions and with the Inhabitants of George's Island'. They were displayed in the ship's lobbies and included cultivating a friendship with the natives by every fair means and to treat them with humanity; trading to be carried out only through an appointed person; the loss of firearms or working tools to be deducted from pay and private trading in ship's stores to be treated in the same way with the possibility of harsher penalties; no iron or useful articles to be exchanged for anything but provisions. These regulations seemed very restrictive to sailors set down in this tropic paradise after such a long voyage, but Cook had taken note of both Lord Morton's hints on how primitive peoples should be treated and Samuel Wallis' journal descriptions of unruly behaviour by his crew. Cook knew rules were essential for law and order and he was also aware they would need to be policed most carefully.

Cook and Banks went ashore expecting to find things very much as they were described by Wallis, but it was a different scene that greeted them. Many of the houses were deserted or in ruins; the queen's house where Captain Wallis and his crew had been entertained by the *arioi*, the Tahitian organizers of religious ritual, was no longer in existence, nor was the queen herself to be found. It seemed there had been some civil disruption and much of the population had moved away from the area. This was confirmed the next day when a fleet of canoes came around the reef from the west, their occupants offering friendly greetings and bringing some hogs they wanted to barter for hatchets, when the previous going rate had been a few iron spikes.

Priest figure from the Society Islands. After Parkinson.

At a welcoming feast given by the Tahitians Banks noticed among the crowd in the flickering firelight 'a pretty young girl with fire in her eyes', but before he could take advantage of the enticing signs, Monkhouse and Solander suddenly discovered that their pockets had been picked. The realization prompted the gentlemen to stand up abruptly, knocking the butt of a gun, which fell over, causing the native hosts to disperse in fright, 'among whom was my pretty girl', according to a disappointed Banks. The stolen spy-glass and snuff-box were returned but Banks' hopes of an exciting night with an obviously willing Tahitian beauty were dashed.

Cook was not satisfied with the *Endeavour*'s anchorage in Matavai Bay and he thought there might be an improved mooring to the west where the canoes had come from, so he bent the sails and moved around the point, but could find no better position. It was decided to stay where they had originally arrived, close to where a temporary fort would be erected to house the observatory. On the fourth day the outlines for the structure were traced on the sandy shore, tents were set up and sentries posted to guard the construction tools with orders not to allow any of the natives anywhere near. That afternoon Cook and Banks together with Parkinson and Green and a large group of amused Tahitians took an excursion into the country at the back of the beach. It was good to be walking on land again, stretching their legs and enjoying the beauty of the place: the brilliant green foliage of the trees and vines, a colourful profusion of flowers and of birds, the sound of surf booming on the reef in the distance and an atmosphere heavy with tropical perfumes. Banks enjoyed some sport with his fowling piece, bagging three ducks with one shot. In the meantime, a marine guarding the encampment on the beach allowed some of the curious natives to approach too close and, while momentarily distracted, the musket was wrenched from his grasp and the assailant tried to stab him with its bayonet before running away with the weapon. The midshipman in charge hastily gave the order to fire on the intruders and the marines eagerly obeyed, keen to get some action. They behaved as if they, too, were shooting at wild ducks, killing the man who had run off with the gun and wounding several others. The musket was successfully carried away by other natives into the trees. The gentlemen's idyllic stroll in the woods was brought to an abrupt halt by these distant sounds of firing. They hurried back to the shore, catching a glimpse of frightened natives rushing through the undergrowth, then noticing blood staining the dark volcanic sand. Cook's good intentions of cultivating a friendship with the natives had begun badly, his overtures of peace now being thrown into doubt. He ordered the ship to be brought nearer the shore so that her guns could cover this part of the bay and particularly where the fort was to be erected. Joseph Banks was anxious to establish good relations because they were the key to studying a whole new world of primitive customs and nature and there would be much to note, interpret and experience over the next few weeks. He managed to find an old man who agreed to mediate in the affair, and a group of natives came to the tents holding plantain branches as a sign of peace and clapping their hands to their chests, repeating 'taio' in friendship. Polynesians and Europeans sat down together in apparent harmony, although it was an uneasy meeting with the vivid memories of the shooting so fresh in everyone's minds. Sydney Parkinson, who was horrified that the marines had opened fire with such glee on unarmed natives, came to the conclusion that 'resentment with them is a short-lived passion'.

The general area of the bay was obviously not secure enough and Cook ordered the tents to be struck and taken on board the *Endeavour* for the night together with all the tools. Another site would need to be chosen for the fort. Alexander Buchan, who had seemed in good health since his epileptic fit at Tierra del Fuego and worked industriously at his drawings during the long voyage across the Pacific, was suddenly seized again with convulsions, but this time they were prolonged and more serious. He lay in a coma for 24 hours and then died. The shooting incident had upset relations with the natives and Banks was sensitive to the fact that a European burial on their soil might offend local customs and upset them even more. Cook's solution was to take the pinnace and the longboat out through the gap in the reef for the funeral and, with the cloud-capped peaks of the island of Moorea looming in the distance, commit poor Buchan's tortured body to the blue waters, 'with all the decency the circumstance of the place would admit of'. Banks was beginning to appreciate what treasures were all around and wrote of his draughtsman's death, mourning the loss of Buchan's services rather more than the man: 'I sincerely regret him as an ingenious and good young man, but his loss to me is irretrievable. My airy dreams of entertaining my friends in England with the scenes I am to see here are vanished. No account of the figures and dresses of men can be satisfactory unless illustrated by figures: had Providence spared him a month longer what an advantage it would have been to my undertaking, but I must submit'. It was Sydney Parkinson who would now have to accept an increasing burden of work in depicting everything from anthropology to botany: his eyes and hands were the only trained means of recording the remaining details of Joseph Banks' grand tour.

To well-read Europeans in the late eighteenth century the natives of Otaheite, as it was now called, confirmed current theories about the noble savage. These were found in philsophical writings, such as those of Jean-Jacques Rousseau and the more popular forms of the day like the French romance *Télémaque*, which had been translated into English by Dr John Hawkesworth. Voyagers to the South Seas were being conditioned to expect neo-classical societies, differing from each other only in their degree of primitiveness. In Tahiti, the innocent natives, far removed from the taints of the civilized world, were imagined as living out their lives in a benign and bounteous existence akin to the ideals of Greek and Roman antiquity. There was no lack of food, for as Banks observed: 'Scarcely can it be said that they earn their bread by the sweat of their brow when their chief sustenance, breadfruit, is procured with no more trouble than that of climbing a tree and pulling it down'. Personal cleanliness abounded and the custom of smearing their bodies with coconut oil was only a mild irritant to the sensitive noses of visitors: 'These people are free from all smells of mortality', Banks noted, 'and surely rancid as their oil is it must be preferred to the odiferous perfume of toe and armpit so prevalent in Europe'. The natives could be seen to live in perfect harmony with both nature and their gods, proof enough that mankind, left to its own devices, was inherently good and confirmation that the evils of so-called civilized societies were a deviation from the natural state. Cook's men, like those of Bougainville and Wallace before them, were unaware of the more complex nature of the island's society, and tended to overlook its darker side of war and human sacrifice, including infanticide and cannibalism. Philibert Commerson, the botanist aboard Bougainville's ship, *La Boudeuse*, in 1768, even described

Tahiti as 'the one spot upon the earth's surface without either vices, prejudices, wants or dissensions'.

Banks gave Greek nicknames to influential Tahitians who would be key people in maintaining friendly relations during the stay. Lycurgus was the eldest son of one of the leading *arioi*, named 'from the justice he executed on his offending subjects', Hercules, 'from the large size of his body', was a powerful chief in the western part of the island whose real identity was Tuteha, and Ajax was so named because of 'his grim countenance'. Thus Banks, at least, seemed to regard this society as some sort of antipodean Athens. He could not know that Bougainville had already named this place New Cythera after the Greek island where Aphrodite, goddess of love and beauty, patroness of prostitutes, rose from the sea.

Now the men of the *Endeavour* set about the main business of the voyage: the erection of the fort and observatory. Earth banks were raised on the site finally chosen, with ditches at either end. A line of barrels filled with sand on which were mounted two of the ship's four-pounder guns faced the river, and on the side overlooking the bay a wooden palisade was constructed surmounted by six swivel guns. The compound contained the observatory, armourer's forge, a cook's oven, quarters under canvas for the officers and gentlemen, latrine — or 'necessary house', as it was delicately named — and outside the walls were tents for the sailmaker and cooper. With accommodation for half the ship's complement and guarded by the marines, what was now called Fort Venus must have seemed impregnable. It was also uncomfortable in the frequently windy conditions when the sand, on which it was built, flew everywhere, permeating the sleeping quarters and the food. Cook considered himself 'perfectly secure from anything these people could attempt'. But he was mistaken. The security was soon tested and found wanting when the heavy brass quadrant, on which the forthcoming observation depended, was stolen. As they were carrying no replacement, the captain and the distraught astronomer turned to Banks for help because of his obvious flair for making friendly contact with the natives. Perhaps he could find out who had stolen the precious equipment and persuade those responsible to return it, otherwise the principal purpose of the voyage could not be achieved. Banks set out with Charles Green and Jonathan Monkhouse in search of a local chief who, it was rumoured, knew about the incident. He was found up in the hills and persuaded to assist in recovering the irreplaceable instrument. When it was finally discovered, they saw it had been broken down into many pieces, although they were all retrieved except for part of the wooden tripod. Fortunately Herman Spöring was adept at repairing such things and with the aid of a set of watchmaker's tools painstakingly re-assembled the quadrant.

Banks, with his winning ways, was appointed by Cook to be the principal trade negotiator in a business where the daily exchange rates fluctuated wildly: sometimes six coconuts could be obtained for an amber-coloured bead, ten for a white one. But it was a substance foreign to their culture that the Tahitians prized most: metal. At first an axe could buy a pig, and the ship carried a few of these for barter. But as the animals grew scarcer, their values rose. It was a far cry from when the *Dolphin's* men two years before could buy a native woman for a nail a night, with the ship almost coming apart at the seams from the rush to gouge out every available piece of metal. When there were no nails left they had stolen lead and cut it into the shape

A native of Tahiti. After Parkinson.

of nails. Under Banks' supervision, a lively market sprang up outside the fortifications where the natives bartered provisions and fruit for metal implements and trinkets.

Contacts between natives and visitors now flowered. At first, the seamen, whose sexual horizons rarely extended beyond dockside prostitutes, were taken aback, even shocked, by the permissiveness of Tahitian women. Everyone had been starved of female companionship for at least eight months and now they faced the ultimate frustration. As Banks explained: 'The ladies showed us all kinds of civilities our situation could admit of, but as there were no places of retirement, the houses being entirely without walls, we had not an opportunity to put their politeness to every test'. These Europeans balked at public sex and it was the ultimate titillation to have willing beauties point invitingly to mats on the floor, yet be unable to accept their charms because of a white taboo.

However, it did not take long for furtive pairings to begin, and some of the men started to make guarded references to the liaisons in their private journals. The master, Robert Molyneux, admitted: 'The women began to have a share in our friendship which is by no means platonic', while his mate, Francis Wilkinson, who nursed many happy memories of these native women from his previous visit, was delighted to find that little had changed: 'We find the women of this island to be very kind in all respects as usual when we were here in the *Dolphin*'. Banks revelled in the Tahitians' frequent uninhibited displays, particularly the dances which, in fact, had a strong religious content. Their movements appeared to him as delightfully indecent actions accompanied by provocative words. Sexual initiations also took place in public involving youths performing intercourse with young girls whose actions were assisted by advice and encouragement from the women. Many of the British onlookers noted that most of the girls, in spite of their tender years, seemed to need little instruction to enjoy their initial experience of copulation.

Only a handful of the *Endeavour*'s men were not actively interested in this aspect of Tahitian life — Cook, Solander and Parkinson among them. Banks, as always, lived life to the full in experiencing all aspects of their culture, including sex, ceremony and food. He was invited to taste a roast leg of dog at a feast and enjoyed it so much that he insisted on having more. A live animal was bought for him and handed over to some natives who agreed to slaughter and dress it and then cook the meat in local style. Banks watched with fascination as the head was pushed right down to the pit of its stomach until it suffocated and then the hair was singed off by turning the carcass on a bed of hot stones. The skin was scraped clean with a coconut shell and pieces of sharp coral. The belly was split open and entrails spilled out on hot stones and quickly broiled, serving as a sort of *hors-d'oeuvre* well before the main meal. The dog's carcass was then placed in an oven of hot stones and covered with breadfruit leaves. The liver, heart and lungs were put on top and a coconut shell brimming with the animal's blood poured over them, before being covered with more leaves and the oven sealed with firmly-packed earth. A cooking time of four hours was allowed and then the delicacy was served for supper. Sydney Parkinson nibbled nervously at a bit of the flesh and thought it tasted like coarse beef, but with a strong and disagreeable smell in spite of the dog being vegetarian. Cook, Banks and Solander judged it excellent, claiming it as the sweetest meat they had ever tasted, although none of the other *Endeavour* men would have anything to do with it. Many of them were to acquire a taste for another curious dish that was loathed by the natives as much as roast dog was disliked by the crew. Vegetarian rats were common around Fort Venus at night and they became a camp delicacy fried for breakfast.

Cook's discipline in Tahiti was intended to be even-handed, favouring neither native nor white in an attempt to maintain harmony between the two races. He issued instructions to his men on their first half-holiday that 'they should in all things behave as if he was present'. A few days later Lycurgus contacted Banks and demanded that he accompany him to a house where an argument had broken out between a member of the *Endeavour*'s crew and a native woman. It had started when the ship's butcher, Henry Jeffs, tried to obtain a stone axe in return for a hatchet and a nail. The woman refused the offer and he had hurled the nail at her, snatching the native axe and threatening murder if she protested further.

When Cook learned of this he was furious that one of his crew had violated the trading code, threatening the carefully nurtured harmony. Jeffs would have a flogging for the offence as an example to others that rules were to be obeyed. Arrangements were made to perform the punishment ashore, but then it was thought inappropriate to dispense British justice on this foreign soil for the same reason that Buchan's burial was carried out at sea. The flogging took place on the deck of the *Endeavour* in Matavai Bay with Lycurgus and his wife invited on board to view the spectacle. They stood quietly as Jeffs was stripped to the waist and tied to the rigging, but as soon as the first blow was ministered they burst out in tears and pleaded for the brutality to cease. It was a request that Cook had to reject as the offender must be seen to receive his full complement of twelve lashes.

Banks had carefully read the accounts of Captain Wallis' visit and was anxious to meet one of the personalities whose name cropped up many times in the journal: the powerful Queen Oberea, or Purea as it was written. In 1767 she was a dominant force in the Matavai Bay area but it was obvious now that circumstances had changed dramatically with her power attenuated and most symbols of her reign dismantled. She was, in fact, sent into exile after a bloody civil uprising and eventually allowed back to this district with diminished influence. She retained, however, the almost legendary sexual powers which, from the moment of their first meeting, haunted Joseph Banks for the rest of his life. The European view of this woman would become distorted, although the basic facts were undeniable, so that Purea became a symbol of all that was desirable and available in Tahitian womanhood. She arrived unannounced at the fort one morning in the company of her priestly adviser, Tupaia, with a present of cloth. From the deference shown her, she was still a person to be reckoned with. Banks estimated she must be middle-aged, she was also rather plump and taller than her companions. He noted: 'She appeared to be about forty, tall and very lusty, her skin white and her eyes full of meaning'. He guessed Purea had been very handsome when young, but that was a long time ago, although Banks would soon discover that an enthusiastic sexual appetite had scarcely been dulled by the years. Her husband was not present these days and she enjoyed the company of much younger men, just as he was reputed to prefer younger women.

Purea invited Banks to visit her house and on arriving with an air of anticipation he was surprised to find her in bed with 'a handsome lusty young man of about twenty-five, whose name was Obadee', her 'gallant'. Clearly no secret was made of their relationship. Without embarrassment Purea got out of bed, dressed herself and placed a length of cloth over the visitor's shoulder as a sign that he would be next to share her favours. But the recent matter of the stolen quadrant had assumed serious proportions and Banks had to force himself to display a coolly diplomatic approach.

The next encounter with Purea was concerned with smoothing ruffled sensibilities over the taking of the quadrant. The chief, Tuteha, indicated he had been deeply hurt by his undignified detention while the *Endeavour*'s search party was away in the hills and Cook realized that an attempt must be made to mollify him in the interests of good relations. With Cook and Solander he visited the chief's house, where they were greeted rather icily at first, but after presenting gifts of a hatchet and a broadcloth cape, the atmosphere began to improve. Purea appeared and bustled Banks off to one of the nearby houses 'where I could be free of the suffocating heat occasioned by so large a crowd of people as were gathered around us'. This was no moment for gallantry, however, because entertainment was planned for the visitors by the *arioi*, including a wrestling match followed by a feast of roast pig and breadfruit. Banks' sexual curiosity about Purea was over-ridden by an even stronger desire for food, with his stomach 'sufficiently prepared for the repast'. Tuteha was not going to allow his wounded pride to recover quite so easily because, in a rapid change of mood, he announced the desire to go aboard the *Endeavour* to eat. Cook was anxious to placate the chief and agreed to one of the roast animals being taken in a boat across the four miles of lagoon to where the ship was anchored. By this stage Banks was ravenous and grumbled that the food was 'growing cold under our noses before he would give it to us'.

The ever-observant Sydney Parkinson was restrained from tasting the sins of the flesh by his religious convictions and this gave him a detachment which allowed close observation of the natives without any emotional overtones. He noted that polygamy was not allowed, but married women had such little sense of modesty that their husbands permitted the visitors to take any liberty with their wives, except intercourse. Most of the ship's company procured temporary wives of their own for the duration of the *Endeavour*'s stay and Parkinson considered this scandalous: 'As if a change of place altered the moral turpitude of fornication'. It was a grave dilemma for him that European sin could be regarded as simple gratification else-where and chastity among women should be prized only in certain parts of the world — certainly not here.

Purea's determination to take Banks as her lover was overwhelming. She decided to dispense with her regular young Tahitian bedmate, who was so put out over the separation that he began to hang around Fort Venus wearing a doleful expression. On one crowded night Banks found himself with the rejected young man, Purea and her beautiful young girl attendant all sleeping in his tent. From that moment he rarely spent a night alone, although it was not always accompanied by unalloyed passion.

At the end of May Cook, Banks and Solander went on a mission to the west in search of more hogs, the main source of fresh meat for the *Endeavour*'s men, and one which had become increasingly scarce. They were met warmly by the chief and Purea and, as it was late by the time they arrived, it was necessary to obtain 'lodgings' for the night, as Banks expressed it. Purea offered him a bed in her canoe which was drawn up on the shore: 'I acquainted my fellow travellers with my good fortune and wishing them as good took my leave'. Banks undressed until he was naked because it was a very hot and humid night and, being exhausted after the day's journeying, settled down to sleep. Purea had insisted on guarding his clothes for fear of them being stolen; Banks agreed and drifted off to sleep 'with all imaginable tranquility'. A couple of hours later the noise of music woke him and when he looked for his clothes they were nowhere to be found. Banks shook Purea awake demanding his possessions and a search was made by candle-light, with him still naked. His fine waistcoat embroidered with silver frogs was missing as well as a jacket with a pair of pistols and powder and shot in the pockets. A musket was also gone. It was no consolation that this premature awakening was followed by some excellent singing and dancing to the accompaniment of nose flutes and throbbing drums. By daybreak, after a restless night, Tupaia had managed to retrieve the musket and most of Banks' clothes, but the jacket and pistols were nowhere to be found. It was soon revealed that others in the party had fared little better: Cook's stockings were taken from beneath his head while he claimed he was still awake and various members of the expedition lost other articles of clothing. The only fruit of this excursion was the acquisition of one hog.

At the end of May, the time was getting close for the transit of Venus and all energies were concentrated on preparations for the observation. Lord Morton's advice was noted and Cook directed that in addition to the main sighting from Fort Venus, there should be others: at the eastern end of Tahiti and from the island of Moorea, or York Island as Wallis named it, on the horizon across the water from Matavai Bay. Banks would accompany this expedition staffed by John Gore, surgeon Monkhouse and Herman Spöring. Their only worry was the weather, which was annoyingly changeable with fine, clear days followed by overcast conditons and periods of torrential rain. In a special tent at the fort, Shelton's astronomical clock and its gridiron pendulum were placed in a wooden frame with a guard rail around to prevent any accidents that might jolt the delicate mechanism. Twelve feet away stood the observatory containing a journeyman clock and Bird's astronomical quadrant, again in perfect working order. The heavy brass instrument was set on top of a wooden cask filled with wet sand to anchor it firmly to the ground. The two reflecting telescopes constructed by James Short were placed in position and everything was put under guard by the marines with orders to let nobody near the equipment except on specific astronomical business.

Astronomers knew that the planet Venus, revolving round the sun inside the orbit of the earth, very occasionally could be seen passing across the sun's disc as a small black dot. Since the transit was first observed by Jeremiah Horrocks in 1639 and again 40 years later by the Astronomer-Royal, Edmond Halley, it was realized that this phenomenon could be used to calculate the scale of the solar system and the distance of the earth from the sun. Such calculations would not only add to general knowledge of the universe, they also had practical application for navigation, particularly for European nations to which exploration and colonization were of prime political and economic importance. The object of the observations to be taken on 3 June 1769 was, in effect, to fix a fundamental yardstick for all astronomical measurements. As previously mentioned, several countries, including France and Russia had mounted expeditions to gather this information in various parts of the world, both in the northern and southern hemispheres. By setting up observatories at known distances from each other, the track of the transit and its duration would differ according to where it was viewed. A record of those differences from a sufficient number of stations was required for precise results.

Banks had to wait for repairs to the longboat after it was found to be riddled with the teredo worm, making a honeycomb of the lower planks, but then he and John Gore's party were able to proceed while Hicks, Clerke, Pickersgill and midshipman Saunders took the pinnace to the east. Joseph Banks had little interest in the transit and spent his stay on Moorea viewing the impressive landscape with its jagged volcanic peaks, collecting a few plants and trading with the local chief for provisions, including breadfruit, coconuts, a dog and a hog in return for a shirt, an axe and some beads. Later in the day 'three handsome girls' came to their camp and were persuaded to spend the night in his tent, which demonstrated to Banks: 'a proof of confidence which I have not met with upon so short an acquaintance'.

After some apprehension about the day ahead after the night sky was partly obscured by cloud, Saturday, 3 June dawned favourable for the observations, becoming hot and clear. These were ideal conditions for the extended period of the transit, which would last from 9 a.m. until mid-afternoon. At Fort Venus they could see the image of the planet crossing the sun's disc through their telescopes, although vital information about the precise time of the start of the transit and its end could not be assessed. Cook noted that the indistinct image 'very much disturbed the times of the contacts particularly the two internal ones. Dr Solander observed as well as Mr Green and myself, and we differed from one another in observing the times of the contacts much more than could be expected'. The observations by the other two parties were judged to be satisfactory, with Pickersgill stating that his sightings were good 'so that if the observation is not well made it is entirely owing to the observers'. Banks' only disappointment was having to leave his own studies of Venus incarnate in the face of 'the entreaties of our fair companions who persuaded us much to stay'. The day was further marred by the theft of a large number of spikes from the ship's stores and carpenter's mate Archibald Wolfe's flagrant breach of regulations resulted in the severest penalty thus far imposed by the captain: 20 lashes.

James Cook was in no great hurry to leave Tahiti once his main scientific mission was completed, although the season was advancing and local food supplies were in short supply. There were now southern latitudes to explore in search of the elusive continent, but his ship needed to be in peak condition before departing into the unknown. Repairs and maintenance had been carried out on the *Endeavour* ever since they had arrived two months before, although, with half the crew living ashore making preparations for the transit, work had progressed slowly. The vessel was heeled over where she lay in Matavai Bay to have her hull cleared of marine growths before a coat of protective pitch and brimstone could be applied. Timbers above the waterline were caulked and painted, powder was dried, the rigging thoroughly inspected and repaired where there was chafing, masts and spars received gleaming coats of varnish, and sails were dried and repaired. Every part of the ship was cleaned and fumigated in preparation for the onward voyage. But the *Endeavour*'s clean bill of health did not extend to many of the crew. There was now venereal disease among the men after their association with Tahitian women, which mystified Cook and surgeon Monkhouse because Wallis' example of having everyone inspected for signs of the disease before reaching the island had been carefully followed. Wallis' men had left without ill effects and Cook's own complement, with the exception of one man who had no further sexual contacts, were unaffected when they arrived. By the time of the transit about a third of the *Endeavour*'s crew had various stages of gonorrhoea, for which no cure was known at that time. It led Cook to enquire if there had been other visitors after Wallis and before themselves and he learned of the visit of a foreign ship, like his own, perhaps a year before. The natives who revealed this information were shown pictures of the flags of European nations and after much deliberation among themselves the colours of Spain were identified as those that had flown over Matavai Bay after the departure of the *Dolphin*. This was surprising and frustrating news because there was no way of having it confirmed or denied until the *Endeavour* came into contact with European civilization once again. Cook presumed it must have been the sailors from this mystery ship who had brought venereal disease to the island.

Banks welcomed the prospect of an extended stay. The duties of diplomat and chief barterer amused him, giving ample opportunity to study native customs and experience some of the intimate details of their lives. This diversion, however, meant that his own botanical collecting suffered. Even the industrious Parkinson spent considerable time away from his paper and paints to study the native way of life for his journal and make a vocabulary of Tahitian words. Only the faithful Solander kept alive the business of natural history.

James Mario Matra was also busy filling page after page of his diary with information about life in Tahiti. He noted that the islanders' complexions were brown, but much lighter than that of the natives of America, and he saw some red hair, although it was commonly black and straight. He thought the Tahitians were less dictated to by custom and fashion in their dressing than Europeans, depending on fancy, caprice or the state of the weather. The men's pubic parts were kept covered during the day and the women 'thought it highly ornamental to enfold their pubes with many windings of cloth which was drawn so close about the middle and round the upper part of the thighs, that walking was difficult for them'. The men, who were usually bearded, were circumcized: 'from notions of cleanliness' and their term of reproach to the uncircumcized was not able to be logged, for the sake of decency. Matra thought the women were well-proportioned, sprightly and lascivious, 'not esteeming continence as a virtue, since almost every one of our crew procured temporary wives among them, who were eagerly retained during our stay'. He noted no particular form of divine worship but observed that in eating they frequently cut off a small portion of their food and deposited it in a secret place as an offering to 'Maw-we'. Their eating habits he thought immoderate because they tended to swallow several large mouthfuls at once.

The natives continued to pilfer nails from the camp and eventually Cook was forced to set an example when a series of thefts, including firearms, culminated in the loss of an iron oven rake. The captain immediately ordered the impounding of a large group of double canoes which arrived near the ship. Some had come from another island, but ignoring this, he threatened to burn all 27 craft if the missing objects were not returned speedily. The rake reappeared before noon, but nothing else. The canoes remained in custody. In retaliation, Tuteha informed Banks that he would no longer supply provisions to the visitors and it became a ridiculous stalemate, with neither side prepared to give an inch until four days later when Purea and several of her attendants arrived in a canoe laden with plantains, breadfruit and a hog, but no news of the other missing objects. Purea explained that her lover had probably taken them and then disappeared. Cook continued to show his disapproval by refusing to accept gifts and Banks added his own rebuff by denying Purea access to his tent. She returned the next day and the routine was repeated. This time Cook softened his attitude a little by accepting some fruit while two young girl attendants looked invitingly around and were soon taken up by William Monkhouse and Zachary Hicks, staying with the officers for most of the day until both girls decided to move into Banks' tent. Monkhouse was extremely jealous of this cupidity and he had a violent argument with Banks which was so animated that Parkinson, who watched the proceedings, feared it must end in a duel. However, the two men knew each other too well from times past in Newfoundland, when Monkhouse was surgeon on the *Niger*, to engage in such extremes. In the meantime, Purea and her permissive companions had returned to their canoe and, after the threat of violence subsided, Banks joined them and enjoyed the night in their company. After a week Cook was forced to allow the impounded canoes to go free because the owners obviously had nothing to do with the thefts. By this time their catches had been exposed to the hot sun every day and the whole bay reeked of rotting fish.

The sublime situation in which most of the crew had found themselves for the past three months had taken the edge off the desire to go to sea again. Life there was hard and dangerous and the sojourn in Tahiti had been more like a holiday with gentlemen, sailors and officers alike revelling in the beautiful climate, enjoying the fruits of the soil and the charms of the women. This, combined with the uncertainty of the voyage ahead, prompted schemes of desertion. Not a few of the *Endeavour's* crew thought the severe penalties for mutiny, or at least desertion, might be worth risking for the delights of living permanently in this earthly paradise. But these essentially private fantasies did not develop into any organized effort. As the day of departure drew closer, the local attractions proved too tempting for only two of the men. Clement Webb and Samuel Gibson, both marine privates, headed for the mountains of the interior with their women, who they termed 'wives', determined to retain their menages. Cook was faced with the decision of whether to leave them and journey on with two marines short or try to recover the men with the assistance of natives. He decided he could not afford to have only nine marines on the next stage of the voyage, which was potentially hazardous, and initiated a determined effort to retrieve the deserters. Cook knew this could only happen quickly by using the strong-arm tactic of seizing a number of Tahitians as hostages until the missing men were returned. Purea and Tuteha were among those held in custody while midshipman Jonathan Monkhouse and John Trusslove, the corporal of marines, set out in search of the missing crewmen. Webb was soon found and returned, but in apprehending him the *Endeavour's* men were captured by the natives and treated very roughly. More marines were despatched in the longboat and eventually came back with Monkhouse, Trusslove and the deserting Gibson.

Throughout their stay the visitors had been asked by several Tahitians to take them as passengers when the *Endeavour* departed. Cook was opposed to this but Banks, thinking of his return home, was attracted to the idea of acquiring a noble savage as a human pet. The priest Tupaia was one of the most persistent in his demand to travel and Banks became his champion, scheming that he might be kept 'as a curiosity as well as some of my neighbours do lions and tigers at a larger expense than he will probably ever put me to. The amusement I shall have in his future conversation and the benefit he will be of to this ship, as well as what he may be if another should be sent into these seas, will I think fully repay me'. The captain finally agreed, persuaded mainly by the suggestion that Tupaia could be useful as an interpreter and navigator for the other islands in his region and possibly throughout the rest of the Pacific. He was allowed to join the *Endeavour* accompanied by a young boy named Taieto, who was a servant and perhaps his son.

On the morning of 13 July Banks invited Purea together with several other friends on board the ship to say goodbye. The farewells were tender and yet restrained, although all around were canoes filled with natives wailing lamentations. Banks sceptically regarded the cacophony as affectation rather than real grief at their departure, but noticed that Tupaia was genuinely moved by leaving his people. A light easterly breeze filled the *Endeavour*'s sails for the first time in three months as she gradually gathered speed, butting her way through the gap in the reef, heading for the open sea.

5
Tahiti — New Zealand
1769−1770

They are ill discoverers that think there is no land,
when they can see nothing but sea.

Francis Bacon (1561−1626)
The Advancement of Learning

COOK knew there were several more Polynesian islands to the west and he
intended cruising them before heading south in search of the elusive Terra
Australis Incognita. For a month it was mostly pleasant, relaxed sailing
through the Society Islands, named in honour of the voyage's patron, with
landings for the botanists to add to their collections, further opportunity to
observe local ceremonies, negotiate fresh provisions at reasonable rates, and
be accommodated by compliant women. Tupaia proved himself a useful
guide, advising on local customs and politics as well as interpreting the
regional dialects, although his prognostications had a certain transparency.
The first visit was to Huahine, where the ship was carried before a fine
breeze which the Tahitian priest claimed to have conjured up through the
good auspices of a friendly deity. It was easy to see through his wiles, as
Banks found out: 'Our Indian often prayed to Tare for a wind', he noted,
'and as often boasted to me of the success of his prayers, which I plainly
saw to never begin 'till he saw a breeze so near the ship that it generally
reached her before his prayers were finished'. The Huahine natives were less
gregarious than their fellows on Tahiti and seemed to spend much of the
time in prayer rituals. After giving them token trinkets in honour of their
gods, a pig was offered in return for the visitors' own supreme being which
Banks joked 'will certainly be our bellies'.

The time for exploration had come, and in early August Banks wrote a
nonchalant note in his diary: 'Launched into the ocean in search of what
chance and Tupaia might direct us to'. A close watch was kept on the
southern horizon for any sight of land and, although a couple of long cloud
banks looked like coasts and Tupaia even gave them a name, the broad
Pacific swells and a lack of currents told seasoned sailors there was unlikely
to be anything but open ocean for countless leagues. By the beginning of
September in rough, grey seas and with bitingly cold winds, Cook estimated they were beyond the latitude of 40 degrees south ordered by the
Admiralty. The *Endeavour*'s sails and rigging were taking a battering from

the heavy weather and he decided to take her north again into more temperate conditions, heading north-west right up to 29 degrees and then turning back to 38 degrees south, with still no signs of land.

Meanwhile the gentlemen worked in the great cabin day by day in a cosy routine that Banks wished could be seen, with the aid of a magical spying glass, by their friends in England. A peep at their situation, he said, would reveal Solander sitting at the cabin table classifying, himself at the bureau making journal entries; between them hung a large bunch of seaweed for study and on the table lay a barnacle-encrusted piece of wood taken from the ocean in case it should indicate the proximity of land. The magical glass would also show a pause and their lips moving to discuss one of the main topics of conversation: what would be found in the territory soon to reveal itself over the horizon? Parkinson sketched and painted, his materials propped up to stop them from sliding off the table, while Spöring jotted information in the notebooks with his clear, precise hand. Hour after hour they sat and worked, anxious to get everything from the Society Islands recorded before the next landfall, wherever that might be. Sydney Parkinson completed 114 watercolours of the plants of the Society Islands from the hundreds collected, the first examples of Pacific islands flora systematically classified by Europeans. The brilliant hibiscuses were to become a symbol of Polynesia and together with the legumes, convolvulus, flowering shrubs and fruit-bearing trees would provide the most colourful section of the *Florilegium*, with 89 plates.

There were several more false sightings of what Banks described as 'our old enemy Cape Flyaway', which dissolved into clouds as soon as hopes were raised. Schools of porpoises circled the ship, leaping out of the water, while skuas with brown bodies and wings, as big as crows, circled overhead. Discussions would continue in the great cabin about the best prospects of finding the great southern continent, which Banks still believed to exist. Eighteenth-century Europe was attracted to the concept of a vast territory in the southern hemisphere to balance the land masses of the north. The French in particular advanced several theories of equilibrium and the obvious need for a balance in the South Seas, which were known to cover more than a quarter of the earth's surface. The land mass envisaged by writers such as Buffon and Maupertuis in the mid-1700s embraced an area as large as Europe, Asia and Africa combined. The most persuasive argument came from Charles de Brosses, a judge and politician in the Burgundian city of Dijon at a time when the French provinces were alive with intellectual activity. De Brosses collected accounts of all the voyages made to the South Seas by the Spanish, Portuguese, Dutch and British and had them translated into French. His analysis of these journeys resulted in a scheme for searching for the missing continent and suggestions about how it should be colonized and developed by the French. His book was published in 1756 in two quarto volumes, *Histoire des Navigations aux Terres Australes*. Its English edition of December 1766 was one of the most carefully studied in the *Endeavour*'s library. France lost most of her colonial empire during the Seven Years War that began the year de Brosses originally published, including possessions in Canada and India. The British government was aware of Bougainville's 'secret' mission in late 1763 to the Falkland Islands, which were seen as a strategic base for South American contact and a stepping-stone to the Pacific.

Given the dwindling supply of fresh provisions and water, land could not be discovered soon enough. Only a few weeks after leaving Polynesia the hogs and fowls taken on board began to die of the cold and for want of proper food. Banks considered he was comparatively well-off for 'refreshments', with the ship's beef and pork still in excellent condition as well as the peas, flour and oatmeal. Surprisingly, even the water remained sweet, 'and has rather more spirit that it had when drunk out of the river at Otaheite'. Wheat was boiled for breakfast two or three times a week and he felt it helped to insulate them from the cold. Of Banks' private supplies, there remained 17 live sheep, 4 or 5 fowls and South Seas hogs, the same number of muscovy ducks and a sow with a litter of piglets. The main problem with the ship's food was the condition of the biscuit, which was a

Pisonia grandis (manuscript name)
Watercolour drawing by Sydney
Parkinson from Huahine in the Society
Islands.

staple of seagoing diet. Banks was exaggerating when he claimed to have seen 'hundreds, nay thousands' of weevils shaken out of a single biscuit, although everyone agreed they were troublesome. The officers were able to have their supply treated in a moderate oven which made them 'all walk off', but this was not possible for the rest of the crew. Banks sympathized with the men. The sight of weevils was disagreeable enough, but their taste was as strong as mustard. Being the complete naturalist, he studied and identified several varieties infesting the ship's biscuit. Scurvy was being kept at bay but venereal disease had taken its hold of the crew after the Society Islands and there were 33 confirmed cases on board. Banks agreed with Cook's opinion that the gonorrhoea and syphilis present in Tahitian society must be the direct result of the visit by the mysterious ship they had heard about. 'When we first discovered this distemper among these people', he wrote with genuine concern, 'we were much alarmed, fearing that we ourselves had, notwithstanding our many precautions, brought it among them'. It was not until later that they learned the 'Spanish' ship was in fact French, commanded by Bougainville, and drew their own conclusions.

Nicholas Young, the surgeon's boy, was at the masthead on 5 October when he sighted land far away on the starboard bow. Banks happened to be on deck at the time and amused himself by watching the reactions of the crew who rushed up from below and quickly convinced themselves of the existence of a coast although nothing could be seen from where he was standing: 'yet there were few who did not plainly see it from the deck till it appeared that they looked five points wrong'. Several hours later, by sunset, everyone could see what was possibly an island, certainly not a cloud bank, thrusting high into the sky. There were hills clothed with trees, although Banks thought the general appearance was not as fruitful as he would have wished. Whatever would be revealed, the young boy received the honour of having the landform named after him: Young Nick's Head. This entry on the chart was, in fact, the first general recognition of the lad's existence. He was about 12 years old and must have been taken on board secretly by one of the seamen at Deptford or Plymouth, because his name only appeared on the muster roll after the *Endeavour* reached Tahiti. Nick's reward for being the first to call land was a gallon of rum, which was consumed as surreptitiously by his colleagues as Nicholas Young himself had materialized on board. What he had sighted was the eastern part of the North Island of New Zealand.

Next morning they found themselves in a wide, open bay without apparent shelter, stretching into low land with hills gradually rising in the distance. Smoke was drifting up from a number of places close to the beach and further away indicating to Banks considerable habitation. He noticed a sort of fence on a hill-top and conjectured that it was a deer park or a field of oxen and sheep. It was, in fact, a Maori fortified village, or *pa*. The *Endeavour* had anchored just off the mouth of a river and now they prepared to set foot on this new land. There were small groups of natives watching from various vantage points and, at first, they seemed to be taking scant notice of the visitors. The yawl with Cook on board was just reaching the shore in search of water when several Maoris, who had been hiding in the bushes, suddenly rushed forward and threatened the party with long wooden spears, or lances, as the British called them. The coxswain in the pinnace a short distance away saw what was happening and quickly fired his musket over their heads, but to little effect: the natives still threatened his captain. The firing was repeated, this time straight into them, and one man fell dead from a shot through the heart. Thus the first Maori blood was spilt by a European, and it proved to be only a prelude. The Maoris were a far more warlike race than the Tahitians. Banks showed no inclination to call them by Greek nicknames, but if he had they would be the Spartans of the South Seas, where the Society Islanders had been the Athenians and the miserable Fuegans the Scythians. The sound of firing brought Banks and his group, who had been collecting nearby, rushing to the scene and he came upon the man slumped lifeless on the beach. Quite clinically he noted the tatoo of spiral lines on one side of the face, the clothing, 'of a manufacture entirely new to us', and the way the man's hair was tied up in a knot on top of his head. The search for fresh water was unsuccessful with only a brackish supply discovered, but some wild ducks were shot for eating and there were a few interesting plant specimens to be collected. Their return to the *Endeavour* that evening was accompanied by sounds of protest from a large gathering of Maoris on the shore. This continued throughout the night and Cook ordered a careful watch to be kept in case of a surprise canoe attack. The first day in this new country had been far from auspicious.

Next morning the captain continued his search for water by landing at a different site to the west of the river and he thought it wise to make a show of strength to the inhabitants by taking out all three boats guarded by marines. There was difficult surf to negotiate followed by a hostile reception from the natives, who were waiting positioned on the far side of the river. As the visitors landed the Maoris broke into a ferocious war dance, waving their weapons in the air and chanting in a threatening manner. Tupaia found he was able to make himself understood by shouting at them in his language and he persuaded about 30 young warriors to swim across the shallow water to meet the *Endeavour*'s men, explaining that their mission was peaceful, wanting only to find fresh water and to replenish depleted supplies by trading in provisions. The young Maoris were very wary of the motives and Tupaia warned the captain that the situation was tense and likely to remain volatile. A few trinkets were handed over but were received with indifference, and there was no offer of goods in return. The main interest centred on the visitors' weapons and several attempts were made to snatch them away; in a scuffle Charles Green lost his hanger, a light sabre, and Cook immediately ordered the thief to be shot at. Banks raised his musket loaded with small shot to the shoulder, took careful aim and fired at the fleeing man's back. It stopped the thief's shouts but not his progress. William Monkhouse, who was closer, fired a ball which brought him down. The other Maoris retreated angrily, the badly wounded being carried away, as Mr Green hastily retrieved the hanger and replaced it in its sheath.

Cook was worried by these events, but the most important consideration at the moment was water, and they took off along the bay in search of it. The surf made another landing impossible that day and it was decided to make a survey of the area instead. While this was being carried out two canoes appeared, both about 30 feet in length, one with oars and the other driven by a matting sail. Cook intercepted them and tried again to make friendly contact but the occupants of the first craft fled for the shore when they saw the *Endeavour*'s boats approaching, and the other canoe was surrounded. Their sail was brought down and the Maoris began to hurl stones and wave their oars in an aggressive fury, which caused such confusion that Cook, Banks and Solander simultaneously fired at them, killing several and wounding others. A full-blooded fight ensued. The result, with native wood and stones ranged against firearms, was inevitable. Three young Maoris attempted to escape by swimming to the shore but were picked up and taken back to the *Endeavour*. At first they were resigned to being killed and stood cowering on the deck waiting for the fatal moment; when that did not eventuate the youths became, according to Cook, 'as cheerful and as merry as if they had been with their own friends'. Tupaia talked with them while they devoured weevil-infested ship's biscuit with apparent enjoyment and were given clothes to wear. They were held on board overnight and, after a huge meal of salt pork and more biscuit, settled down to sleep on the deck to the sounds of a second night of unrest from the shore. The captain and gentlemen made an agonizing appraisal of the day's events. Cook could not justify the killings in the canoe: 'They had given me no just provocation and were wholly innocent of my design'. Banks was less awkward and showed more emotion in expressing his feelings: 'This was the most disagreeable day of my life, black be the mark for it and heaven send that such may never return to embitter future reflections'. Such incidents, as much as they might be regretted for humanitarian reasons, would be the way of life and death during the stay in New Zealand.

Next morning the three Maori youths were put ashore and seemed rather reluctant to leave the British hospitality that had become synonymous with full bellies. Some wood was gathered by the *Endeavour*'s crew on this shore excursion and the botanists managed a trip of about a mile into a swamp, but it gained them no more than three new plants. They had to be content with a total of only 40 species in their collecting boxes from this landfall which, according to Banks, 'is not to be wondered at as we were so little ashore and always upon the same spot'. There were at least 200 armed natives nearby and Cook was anxious to avoid another confrontation, so he ordered everyone back to the ship with the intention of departing early the next morning. They set sail on 11 October without a name being given to their first anchorage in this new country. At first Cook had thought it should be called Endeavour Bay, but so proud an association for such a disappointing place seemed inappropriate and it was soon changed to Poverty Bay because 'it afforded us no one thing that we wanted'. That afternoon, with little headway and caught in a calm on the south side of the bay, several canoes came out to the ship and the natives engaged in active trading and went willingly on board. This time the contact was without incident and showed every sign of friendship, although some of the men were recognized as members of previous war parties. The Maoris invited Banks and Solander to accompany them back to the shore, but Cook would not allow it because he was anxious to be under way in search of a better harbour and fresh water as soon as a breeze sprang up. The aimiable natives left the ship, with Banks hoping that on their next encounter 'we may not there have the same tragedy to act over again as we so lately perpetrated'.

Progress down the coast was very slow and punctuated by the approaches of canoes full of warriors, who were given the occasional burst of grape shot to keep them at bay. The value of having a strong and efficient party of marines on board was by now quite obvious. In mid-October, with the ship in Hawke's Bay (named after the First Lord of the Admiralty), yet another flotilla of canoes approached and in a friendly fashion engaged in trading

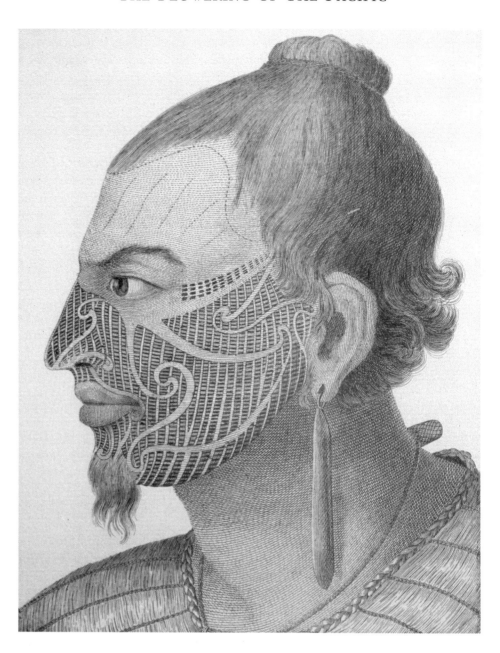

Maori with decorative tattoo. After Parkinson.

dried fish and artifacts for cloth. Cook was particularly interested in acquiring a dog skin cloak and offered a piece of red baize for it. Tupaia's boy Taieto went over the side and was helping to hand the bartered goods up and down when he was suddenly snatched into one of the canoes, which was quickly paddled away towards the shore. The marines on guard saw the incident and immediately opened fire, killing several Maoris. In the confusion Taieto was able to jump overboard and was rescued as the natives made for the river mouth. The attempted capture of the young Tahitian added another colourful name to Cook's charts: the land at the southern end of the bay was called Cape Kidnappers. The *Endeavour* continued southward for another two days, sailing down the rugged coastline with mountainous interiors displaying snowcapped peaks. No good refuges were offered, however, and according to Cook, 'seeing no likelihood of meeting with a harbour and the face of the country visibly altering for the worse', he decided to go about and return north to explore the land beyond their original landfall. The ship turned opposite a point that was named Cape

Turnagain to mark the change of course.

They doubled back on a track that was set further out to sea and when next the land was approached it was in more congenial surroundings, offering plentiful supplies of fresh water and also a profusion of wild celery, which pleased Cook with his continuing obsession about the crew's diet. The people here proved to be more friendly and there was considerable cultivation of crops, although no animals were in evidence except dogs and rats, but plants for the picking grew all around. The Maoris here had one 'necessary house' for every four dwellings, tidy rubbish dumps, and a sense of order that was generally lacking in Tahiti.

Banks and Solander inspected Maori weapons. An old man showed them wooden lances, ten to fourteen feet long, and a *patu* made out of nephrite or greenstone, which Solander agreed was a very handy club, apparently able to split skulls with single blows. The botanists ranged around the bay, collecting many plants and shooting birds, while Spöring and Parkinson sketched. Banks recorded that the plantations grew sweet potatoes, yams and taro in neat rows, estimating their size as up to ten acres and there might have been as much as 200 acres of land under cultivation in the area, tightly fenced with reeds 'so that scarce a mouse could creep through'. The visitors were received graciously and without fear. The women were judged as plain and made themselves more so, in Banks' opinion, by painting their faces with red ochre and oil, which remained wet and shiny on the cheeks and foreheads. There was constant nose-rubbing in the traditional Maori greeting so that Banks and Solander ended up with red noses, making them look rather like circus clowns. Banks thought the women were 'as great coquettes as any Europeans could be and the young ones as skittish as unbroke fillies', but they were unapproachable sexually and the visitors began to yearn for the compliance that had become familiar in Tahiti.

Continuing the journey north, East Cape was passed on the last day of October. There a group of armed Maoris in canoes threatened the ship from a distance, in spite of Cook's hope that word of the *Endeavour*'s presence on the coast had travelled this far and eased the way for friendly contact. Those hopes were quickly dashed when a huge war canoe glided onto the scene, unlike anything they had seen before. There were 60 men on board with 16 paddles a side and a long row of natives in the centre of the vessel from its highly-decorated prow back to the stern. On a signal from a small accompanying canoe, they moved briskly towards the *Endeavour* in what was obviously an attacking manoeuvre as the marines and gunners stood at the ready facing them. The war canoe stopped suddenly but did not retreat until a round of shot was ordered to be fired over them as a warning. A relieved Banks observed, 'They saw it fall and immediately took to their paddles rowing ashore with more haste than I ever saw men, without so much as stopping to breathe till they got out of sight'. Cape Runaway was the name given to this part of the New Zealand coast.

Further threats faced them over the next few days, after which a safe anchorage was discovered in time for an observation of the transit of Mercury across the sun, due on 9 November. Charles Green went ashore with his astronomical telescope after an early breakfast and together with Cook and Hicks carried out the sighting in perfect weather conditions. Meanwhile Banks and Solander botanized 'with our usual good success which could not be doubted in a country so totally new'. Later that morning many natives went out to the ship offering a large catch of mackerel and a few John Dory, enough for several casks of salted fish to be put down for future eating. Some of the canoes were obviously from another part of the country, being larger and better constructed than the rest; their people wore finer clothing and seemed to be of a higher social standing, displaying colourful feathers in the hair. Sydney Parkinson traded a variety of artifacts with a handsome young native who was delighted to received tapa cloth from Tahiti in return. This same young man also had a short cloak made from black and white dog skin which John Gore coveted, offering a large length of tapa for it. The terms were agreed and the tapa was swung over the *Endeavour*'s stern on a rope, but as soon as the young Maori had it in his possession, his companions paddled away as fast as possible, shouting and waving their patus and spears, while Gore was left on the deck without the cloak. A flush of anger came over him at being cheated in this way and he fired at the departing canoe, fatally wounding the young native trader, who had not instigated the sudden departure. Cook was still ashore after the Mercury observations and was furious to learn of the incident when he returned. He told Gore bluntly what a fool he had been, although by the time this senseless death was reported in his journal, the anger was toned down: 'I must own this did not meet with my approbation because I thought the punishment a little too severe for the crime'. He hoped they had been among these people long enough to know how to chastise them for minor misdemeanours without having to take their lives. Cook named the place Mercury Bay to commemorate the least disturbing part of the day.

Maori war canoe in New Zealand. After Parkinson.

If not named after the planet it would have been appropriate to call the area Prosperity in direct contrast to the experience of their first landfall at Poverty Bay. Here, Maori women dived for huge crayfish in the surf among the rocks and many of the succulent crustacea of enormous size were brought to the ship to be boiled in John Thompson's largest cooking pots. A short way up one river extensive oyster beds covered every part of the tidal flats and the *Endeavour*'s longboat returned overloaded with the sweet-tasting shellfish. All along the shore more of the wild celery grew and the natives used the roots of a fern to dry and make a kind of flour. There was plenty of this natural food but little cultivation of the soil and the standard provisions on the ship were getting rather low. On one of their collecting expeditions Banks and Solander managed to shoot 20 shags, which were far from ideal for eating but in the circumstances their oily, fishy taste was acceptable. 'Hunger is certainly most excellent sauce', Banks explained, 'but since our fowls and ducks have gone we find ourselves able to eat any kind of birds (for indeed we throw away none) without even that kind of seasoning'. The prejudices of the crewmen at the start of the voyage had long since disappeared and now, more than a year away from home, hunger led them to try almost any kind of food including a local speciality, hot cockles. Banks had high praise for the local 'lobsters': 'They are certainly the largest and best I have ever eaten'. He was intrigued with another native method of catching the crayfish, by walking among the rocks at low water and feeling underneath with bare feet until one fastened onto a toe. The oysters he judged 'as good as ever came from Colchester and about the same size'. They sailed from Mercury Bay on 15 November and were kept under surveillance from Maori canoes for most of the journey along the coast. 'Come ashore and we will kill you all', was a threat hurled at the *Endeavour* as she passed. Tupaia translated the distant shouting for Banks and Solander and then gave as good as he received: 'While we are at sea you have no manner of business with us', he screamed back, 'The sea is our property as much as yours'. This unexpected outburst of logic took Banks by surprise and, on reflection it confirmed for him the essential goodness of the noble savage: 'Such reasoning from an Indian, who had not the smallest hint from any of us, surprised me, much'. It was quite obvious that Tupaia was going to be a sensation when he reached London.

Harassment from the natives occurred almost every day, met with routine repulsion by musket fire or grape shot. Gradually the *Endeavour* moved north to some of the most beautiful anchorages the men had ever seen: sub-tropical, glowing with greenery and other colours, brilliant under the high sun of the southern spring. 'The country is generally covered with an abundant verdure of grass and trees', Banks noted, 'yet I cannot say that it is productive of so great a variety as many countries I have seen'. The novelty of what was found compensated for the lack of diversity: local herbs, wild celery, cabbage tree and fern roots. The natives' natural diet also included dogs, birds — especially seafowl — fish, sweet potatoes, yams and kumaras. Another encounter with the Maoris began in a predictable way when a crowd gathered in their canoes around the ship, waving spears and chanting, alternately threatening and friendly. All the captain could hope for was have his marines remain vigilant and ready to fire warning shots if the situation deteriorated. A few natives were allowed on board and given presents but while this was happening, others tried to carry off the anchor buoy. Cook ordered several muskets to be discharged as a deterrent but they had little effect and it was not until one of the large guns was primed and fired that the Maoris moved away. Banks was informed by the captain that it was probably now safe to make a landing to collect plants in the area, but no sooner had the party set food in a sandy cove than they found themselves surrounded by hundreds of armed warriors. Banks and Cook remained calm, although expecting to be attacked at any moment. They approached the leaders standing with arms raised ready to strike with their greenstone clubs. The majority of the crew still aboard the *Endeavour* watched with alarm as the confrontation developed. All that could be done from that distance was turn the vessel as quickly as possible so that the large guns might be trained on the beach. The Maoris broke into a war chant and then a wild dance, twisting faces and sticking out their tongues, as the Englishmen stood their ground, knowing that in a sudden rush they must be overwhelmend by the vastly superior numbers. Some natives tried to pull the *Endeavour*'s boat from the shoreline onto dry land and this was the signal for Cook to act; muskets loaded with small shot drove them back in surprise and they were able to be held at bay for about 15 minutes while the ship under the command of Lieutenant Hicks was being manoeuvred. When finally she was in position, guns were fired over the natives' heads with the desired effect of dispersing the crowd as quickly as it had assembled, allowing the shore party to lay down their arms and get on with the peaceful task of collecting wild celery.

A reflective Sydney Parkinson noted in his journal later: 'Had these barbarians acted more in concert, they would have been a formidable enemy, and might have done us much mischief; but they had no kind of order or military discipline among them'. Soon after what might have become a massacre of the British, the natives turned friendly and approachable in a complete reversal of behaviour. Some of them went out to the ship with gifts of large mackerel which served very well for the gentlemen's dinner. This prompted Banks to make a general observation about their behaviour: 'I have often seen a man whose next neighbour was wounded or killed by our shot not give himself the trouble to enquire how or by what means he was hurt'. He thought the Maoris worked themselves into an artificial courage through war dances and chants 'which does not allow them time to think'. The shore party collected their celery while the captain and gentlemen rowed to a nearby cove and set out to climb a hill where they could get a good view of the many surrounding islands and their anchorages which, according to Banks, 'must be as smooth as mill pools as they landlock one another numberless times'. There were many settlements in the vicinity and they had 'certainly seen no place near so populous as this one'. Banks and Solander made their way back to the boat intending to collect plants as they went but there was little to interest them, 'for of all the places we have landed in, this was the only one which did not produce one new vegetable'. The local natives, who had been coaxed into friendliness by the intruders and became 'as tame as we would wish', prepared to trade dried fish in return for paper.

By December the changeable early summer weather settled down enough to allow the *Endeavour*'s departure from the beautiful Bay of Islands. As they neared the northern tip of New Zealand, which Tasman had named Cape Maria van Diemen after the wife of the governor of Batavia, the weather conditions became variable again, with rain, squalls, mists and calms alternating with high winds and choppy seas. On 13 December the

Endeavour was out of sight of land for the first time since reaching this coast, and the wind continued to blow strongly with heavy swells from the west. Life on board was very trying. Most of the sails were in poor condition and the main topsail split in a strong gust, requiring many hours of repairs by the sailmaker, John Ravenhill, and his assistants. On the morning of the fourteenth, high land with a tabletop formation was sighted and appeared to drop away to the south-west. Cook judged this to be the northern extremity of the coast and logged it accordingly as North Cape. There were no signs of habitation and no native canoes. Banks believed with a deep conviction that this coast must be part of the great southern continent they were seeking and the area was assumed to have been thus far devoid of European visitors. It would have come as a huge surprise to James Cook and the Frenchman Jean François Marie de Surville that their ships were at this moment in close proximity to each other. De Surville commanded his *Saint Jean Baptiste* as part of a syndicate of commercial adventurers seeking riches in the Pacific. Their port of departure was Pondicherry on the Indian sub-continent, which they had left in mid-1769, some six months earlier. On 12 December de Surville and his scurvy-ridden crew were on New Zealand's east coast, almost at the same latitude as the *Endeavour* off to the west. De Surville headed north, and what might have been a surprising rendezvous with Cook was averted by the storm that split the *Endeavour*'s topsail and forced the ship out of sight of land for a short period. The *Saint Jean Baptiste* doubled Cape Maria van Diemen with the *Endeavour* ahead by about 60 miles to the north, both crews in ignorance of each other's presence. For de Surville, the adventurer, this voyage would lead to his drowning on the coast of Peru while Cook, the navigator, continued his surveys.

The *Endeavour* beat about the area in variable winds, making very little advance for a week. Then on Christmas Eve, during a rare period of fine weather, a cluster of islands was sighted which had been named by Tasman the Three Kings. It was a nice seasonal coincidence to come across them at this time, although at first there was some doubt about their identity because the outlines did not match the sketch in Dalrymple's book, which showed them with a smoother, less jagged outline. As it was calm, Banks was able to go off in his small boat and shoot several gannets, or Solan Geese as he called them, 'so like European ones that they are hardly distinguishable from them'. Goose pie was the gastronomic highlight of the gentlemen's and officers' Christmas dinner, the second they had celebrated together at sea, and the pie 'was ate with great approbation'. The day was very similar to the previous occasion, only their position on the globe was vastly different. Banks noted that by evening 'all hands were as drunk as our forefathers used to be upon the like occasion'. By the morning of Boxing Day they were heading south off the west coast with many sore heads from the previous day's 'debauch' and running with such high winds that they whipped up, in Cook's words, 'a prodigious high sea' which caused the vessel to be brought to under a reef mainsail. He had been determined to fix the position of Cape Maria van Diemen from the west but the weather was being uncooperative, causing the captain to complain in his journal: 'I cannot help thinking but what will appear a little strange that at this season of the year we should be three weeks in getting ten leagues to the westward and five weeks in getting fifty leagues'. The new year came in with a howling gale and Cook had to ease off the land to avoid risking shipwreck in these 'high rolling seas' from the west. This was frustrating for him because it meant missing landforms and features of the coast which he needed to place on his charts if, as intended, they were to be a comprehensive survey of these unknown shores. It was an audacious action, pitting the little ship, strained and weatherworn, against the prevailing south-westerlies on a lee shore with the constantly changing conditions of the Tasman Sea, but Cook's prudence prevailed: 'I am determined not to come so near again if I can possibly avoid it unless we have very favourable winds indeed'. For the first two weeks of the new year they made their way down the inhospitable 'desert' coast on an uncomfortable journey of squally conditions which never settled into the fine weather expected during the southern summer, but good progress was able to be logged each day. Various landmarks were noted, including a great peak wrapped in clouds with traces of snow on the upper parts. It was named Mount Egmont after a former First Lord of the Admiralty.

Conditions on board the *Endeavour* were becoming increasingly foul and water was running short again. They needed to spend some time in a safe harbour for repairs, maintenance and cleaning. More importantly, it was necessary for the men to stretch their legs on land after being cooped up for such a long time. The previous shore excursions on the east coast had been fraught with danger because of erratic behaviour by the natives and Cook hoped that conditions ahead would be more favourable for a peaceful and

Maori heads. Engraving after Parkinson.

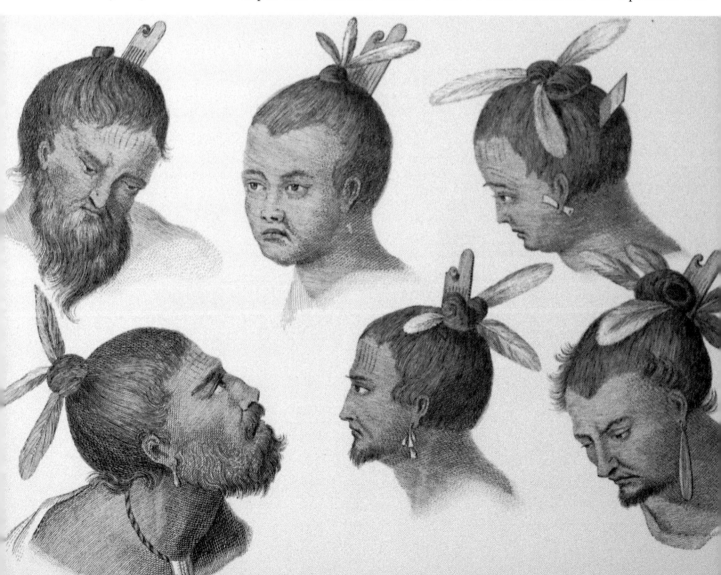

extended stay. On 14 January with land continuing to the horizon, they found themselves in a wide bay. The next day, at the bay's furthest point, they entered a narrow harbour landlocked on three sides. At the entrance were two small islands, one with native fortifications on the top which were crowded with curious onlookers. The *Endeavour* was steered into a little cove and anchored within two cables' length of the shore opposite a rivulet trickling a constant stream of fresh water into the sea. Steep hills towered above, clad with dark green forest and the weather was beginning to act like summer, offering the probability of fine, settled conditions for their stay at this beautiful anchorage, if only the natives proved friendly. There was an abundance of natural food with birds for the shooting and great shoals of fish all around the ship ready to be netted. Mussels and oysters vied for space on every rock, while limitless supplies of wild celery and scurvy grass were to be had on shore. The air was curiously moist, however, permeated with the sickly sweet smell of putrescence, and maggots appeared on birds and fish a few hours after they were caught. The secure mooring rang to the music of bellbirds, greeting the crew in the morning with a metallic tintin-nabulation. 'Their voices', Banks observed, 'were certainly the most me-lodious wild music I have ever heard, almost imitating small bells but with the most tunable silver sound imaginable'. The delightful *aubade* was quickly followed by intense annoyance generated by hordes of minute sandflies whose biting made this apparent paradise quite uncomfortable. The wildlife of the area would come under close scrutiny from the naturalists, with

Idealized view of the New Zealand coast with fortified Maori *pa*. After Parkinson.

much of it falling prey to gun and net, but it was the plants that for the first time in New Zealand would occupy most of their time ashore.

Banks' was attracted to a dress detail on some women he saw in the distance. Bunches of black feathers and rushes covered their heads, making them look twice as tall as they were. He noted to Solander, with a hint of anticipation in his voice, 'These are the most handsome women we have seen on this coast'. But when they got a little nearer he had to revise his opinion. The impressive headdresses had misled him: 'I see not one who is even tolerably handsome'. The first direct contact with the natives was also disappointing: a hail of stones hurled at the ship. But the 300 or so who lived in these parts soon accustomed themselves to the presence of the visitors and became friendly. They appeared to be poorer than their fellows from the north, having rough, undecorated canoes and living off the sea and the land with very little to offer in trade except the fish they caught and some weapons.

After dinner one day the botanists were returning with Cook to their collecting at a cove about a mile from the ship when they saw something in the water that looked like a dead seal. On closer inspection it proved to be the body of a woman, who, from the state of decomposition, must have been dead for some time. They then came across a family who explained to Tupaia that she had been a relative and after dying was buried at sea, but a stone weighing her down became loosened and then dislodged, allowing the corpse to float to the surface. Cook and Banks had thought for some time that the Maoris were natural cannibals and their suspicions were about to be confirmed. When they arrived at the cove the family was preparing the evening meal of a dog baked in an earth oven, very similar to what they had known in Tahiti. There were some baskets nearby containing bones and Banks identified some of them as human, with traces of flesh still evident and gristle on the ends bearing tooth marks. Tupaia was asked to enquire what sort they were.

'The bones of a man', was the reply.

'And have you eaten the flesh?'

'Yes.'

'Have you any of it left?'

'No.'

Apparently a group of enemies had fallen into these peoples' clutches; four were killed and eaten, while three escaped by jumping into the water, but they drowned. The visitors did not let it rest there, however, asking Tupaia to find out why they had not eaten the woman they had seen floating in the cove.

'She was our relation', was the logical reply.

'Who is it that you do eat?'

'Those who are killed in war.'

Banks had noticed there was no cultivation in the area, probably because of insufficient flat land for even a potato garden. He remarked to Cook, 'I suppose they live entirely upon fish, dogs and enemies'. The initial curiosity of the visitors in human bones resulted in a flood of unsolicited offers from the natives, who were quick to see a trading opportunity. They came on board the *Endeavour* sucking them in their mouths and indicating that the flesh was delicious. Four skulls were offered with the brains removed and the eyes missing, but with scalp and hair left intact. Banks bought one to add to his collection of curiosities but there was no interest shown in the grisly objects by other crew members. They were understandably revolted by this trade in human artifacts although Banks explained, 'We were before too well convinced of the existence of such a custom to be surprised, although we were pleased of having so strong a proof of a custom which human nature holds in too great abhorrence to give easy credit to'. He observed that the seamens' faces on learning about cannibalism and seeing the evidence were 'better conceived than described'. Sydney Parkinson, always eager to comment on human frailties, thought it was one of the worst vices, even though the natives seemed to take great pride in their cruelty. After these incidents, evidence of cannibalism was seen everywhere, with human bones scattered through the nearby woods and bailers for canoes constructed from human skulls. The ship's company began referring to the anchorage as Cannibal Bay in the established custom of principal features or incidents giving rise to names for localities.

The definite evidence of cannibalism dampened the desire for sexual contacts with the native women, but one of the crew, who was eager for closer relations, had his pride wounded and returned to the ship in anger after trying to buy the pleasures of a Maori girl in exchange for a piece of cloth. A partner was arranged but, as Banks told it, 'on examination proved to be a boy'. The seaman complained about this in no uncertain terms and a replacement was promised and duly arrived, 'who turned out to be a boy likewise'. When the story got around the ship it led to amused speculation that the Maoris might engage in 'the detestable vice of sodomy' as well as the repulsive custom of cannibalism. This was enough to send young Parkinson into writing numerous admonitory paragraphs about the mistaken ways of primitive peoples. However, Banks thought it more likely that the Maoris were sophisticated enough to play a trick on their visitors as a way of indicating they did not approve of prostitution.

Cook spent most of the three weeks' stay in the bay surveying the immediate area and the nearby coast and unsuccessfully trying to establish where Tasman had lost four of his men, a place he had named Murderers Bay 128 years before. In late January the captain set out with Banks and Solander for the head of the sound in which they were anchored, but after rowing in the pinnace for about 15 miles the end was still nowhere in sight. Leaving the two botanists on the shore to their collecting, Cook climbed the nearest hill with one of his men to see if it was possible to get an elevated panorama. He was rewarded with a magnificent view from a height of about 1,200 feet showing the ocean to the east joined by an open strait to the western sea where the *Endeavour* had sailed. Cook was able to conclude that what he had been sailing around for the last weeks was most probably a large island. He was now standing on the south side of the land masses and needed to confirm whether this was the missing continent, although he suspected not. He later wrote only a modest and restrained account of what he was seeing, although it was one of the great revelations of the voyage. Four days later Banks and Solander joined the captain on a second viewing and saw for themselves the silvery strait stretching from east to west below their feet, although the extremities were lost in the haze of distance. The cautious James Cook was determined to prove the existence of the passage beyond doubt by sailing through and surveying it. On 6 February the *Endeavour* left her snug anchorage; clean, provisioned and prepared to face the rigorous conditions of the southern latitudes for which she would soon be headed. Memories of cannibalism could not be erased from their minds,

Dysoxylum spectabile Watercolour drawing of a common New Zealand tree known to the Maoris as *kohekohe*.

although no reference would appear on the charts. Instead of calling the place Cannibal Bay, as the men knew it, Cook wrote down the more elegant name of Queen Charlotte's Sound, after his king's consort. A course was bent to an opening in the bay, guessing it to be the passage between the northern and southern parts of this land, and the sun went down that evening in a great golden glow as they sat becalmed in the mouth of the strait.

It was not going to be as easy to negotiate this passage as it had looked from above. To the calms was suddenly added a vigorous ebb tide which threatened to drive the vessel on the rocks of a small island. But the momentum was checked by anchoring in 75 fathoms with 150 fathoms of cable out, and by the morning, on the change of the tide, they were able to proceed. Cook needed to be absolutely certain that the northern part was an island because there was a section between their present position and Cape Turnagain that he had not yet surveyed. His own name was given to the strait, proposed by Banks, because the captain would be far too modest to make the suggestion himself. They continued to have contact with the natives on both sides of Cook Strait and noticed a significant difference between them, confirming an early impression about the Maoris of Queen Charlotte's Sound. On the north they were taller and well-built with fine tattoos and wearing good cloth made from flax, while those of the south seemed to be stunted by comparison with little of the sprightliness of their neighbours. By 9 February the *Endeavour* had moved far enough along the coast for those aboard to have a good view of Cape Turnagain, which they had last seen at the beginning of their New Zealand journey four months before. Cook called on deck those few officers who doubted they had

Cook's map of New Zealand reproduced in Parkinson's journal, 1773

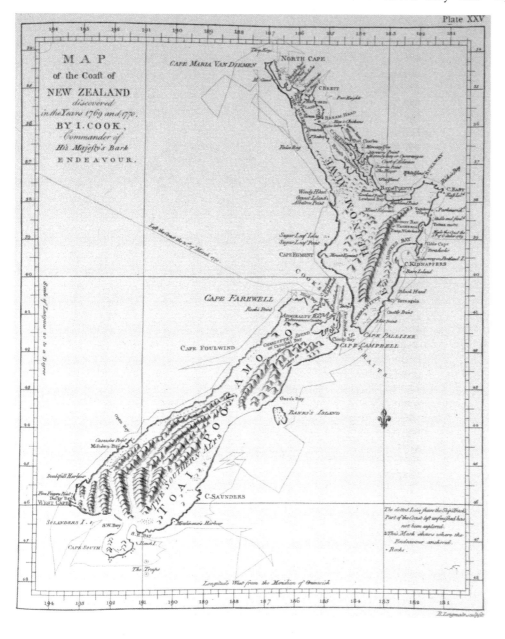

79

circumnavigated an island and when all were in agreement, the ship was turned south with the intention of exploring the east coast of the land on the other side of his strait.

The gentlemen were divided in their opinion on whether the land they were approaching might form part of a great southern continent. Cook was certain it did not, Solander wavered, but Banks held to a conviction that was more than a little coloured by his romantic outlook. Most held no strong opinions, however, and Banks had to admit that only he and one midshipman really cared about the matter. 'The rest', he regretted, 'begin to sigh for roast beef'. The land continuing far to the south raised his hopes a little and he was able to state in a discussion during dinner, 'Again I declare it to be my opinion that a southern continent exists', but by 10 March the question was settled as Banks had to accept, when rounding the southern-most tip of the land, 'the total demolition of our aerial fabric called Continent'. They were now passing some of the most impressive scenery of New Zealand, having turned north-east and then north with land continuous on the starboard bow and penguins swimming with the ship, 'shrieking like geese'. It was a striking landscape with mountains piled on mountains to an amazing height, many of them still topped with snow at this late stage of summer. There were no signs of habitation; no smoke by day or the glow of fires at night. The indented coast seemed to offer some promising harbours but Cook was not prepared to risk going closer to search them out because the Endeavour might be trapped on a lee shore with little hope of easy escape; the possibilities had to be viewed from afar and were awarded such names as Doubtful Harbour and Mistaken Bay. Banks made several requests to go ashore, but Cook thought his persistence 'ill-founded', without considering either the present or future consequences. Banks was left smarting at the rebuff and worrying that there had been so few opportunities for systematic collecting around the New Zealand coast.

The land appeared to Sydney Parkinson 'as wild and romantic as can be conceived'. Snow-capped mountains rose into view one above the other, seemingly from the water's edge. The shoreline was densely wooded with the trees extending back into steep valleys, the higher reaches of which were wreathed in mist. Banks used this period of confinement to collect his thoughts about the country and enter them at length in his journal. He considered the light soil he had inspected was admirable for cultivation and that wooded areas would repay being cleared. The trees impressed him most: 'The straightest, cleanest, and I may say the largest I have ever seen'. They could be of great commercial value to Britain by meeting the demand for the masts and spars of His Majesty's fleet. The place they had named River Thames on the north-west coast of the north island appealed to him most as a site for establishing a European colony and he wrote eloquently of its possibilities: 'A ship as large as ours might be carried several miles up the river, where she would be moored to the trees as safe as alongside a wharf in London River'. The fine timber could furnish plenty of materials for the building of defences, houses or vessels; the river would supply excellent fish and the soil 'make ample returns of any European vegetables grown in it'. The journal entries also included extensive descriptions of native customs, houses, food and weapons as well as a vocabulary compiled with the assistance of Tupaia. While Banks was working at his bureau in the great cabin, with the Endeavour reacting constantly to the broad westerly swells, they again came in sight of Queen Charlotte's Sound, from where they had begun their circumnavigation of the south island six weeks previously. When they had left, the natives had told Cook he could travel around it in just a few days, but it had taken somewhat longer to negotiate this land mass, comparable in size to Britain. Everyone was glad to be back in familiar waters after enduring, as Parkinson so neatly put it, 'the dangers of foul winds and the tedious suspense of many calms'.

The *Endeavour* needed wood and water once again and she was guided into a sheltered bay where these were available. The weather remained gusty and rainy, making for an uncomfortable end to their stay in New Zealand. Four days of cleaning and provisioning were planned to prepare the ship for the next leg of her voyage and in the meantime Cook spent much of his time in the great cabin putting finishing touches to his charts of the entire country's coastline; it has been a remarkable feat of marine surveying in such a short time. The naturalists' work on the table was suspended, after almost annexing it as their own, while the captain and Charles Green entered place names, checked compass bearings and sextant angles against grids of latitude and longitude, and filled in an imagined interior with representations of mountains and valleys. Cook could claim without boasting that few parts of the world were better determined, 'being settled by some hundreds of observations of the sun and moon, and one of the transit of Mercury'. The resulting map of New Zealand was the product of a running survey from the sea with six landing places along the coast in the north and two in the south for a total shore time of less than seven weeks. There were only minor inaccuracies and two incorrect interpretations of land masses, which proved to be not what they seemed from out at sea. The large Bank's Island off the east coast of the south island would be established later as a peninsula, and the southern tip of the country, shown tentatively on Cook's charts as a peninsula was, in fact, a large island. As a companion to his surveys the captain wrote a long and detailed description of each section of the *Endeavour*'s 2,400 mile journey, admitting where closer observation would have resulted in finer detail, but stating categorically that the observations 'will in most places be found to differ not much from the truth'. This was James Cook in his element.

Admiralty Sound, as their final anchorage was named, allowed Banks and Solander to get ashore and make final additions to their collection. Being so close to Queen Charlotte's Sound, where they had stayed for a much longer period, there was little new to be had here, but Banks climbed a hill and was able to find three plants that were unfamiliar to him. They added to the 360 or so species in the first collection of New Zealand flora ever made, and there might have been more, but the total of 44 days the *Endeavour*'s crew had ashore were not exclusively for botany. At the start of their explorations, interference from the natives severely hindered collecting and later the weather was often unfavourable for lengthy expeditions into the country. In addition, the species from the south island were poorly represented, although what they had found was unique to European eyes and they had made their observations during the best season for studying plants, their flowering period. Despite the limitations of this part of the voyage, New Zealand would provide the second largest section of the *Florilegium*: 183 plates.

It was now the southern autumn and time to be moving off on the next leg of the journey. With the charting of New Zealand Cook felt he had fulfilled his obligations to the Admiralty and was now free to take his ship

Plate XIV.

View of the North Side of the Entrance into Poverty Bay, & Morai Island, in New-Zealand. 1. Young Nick's Head. 2. Morai Island.

S. Parkinson del.   R. B. Godfrey Sc.

View of another Side of the Entrance into the said Bay.

View of the north side of the entrance into Poverty Bay, New Zealand. Engraving after Parkinson.

and her passengers home. According to his masters in London that could be either round the Cape of Good Hope, Cape Horn or 'as from circumstances you may judge the most eligible way'. Cook had already made up his own mind to proceed by way of Batavia and the Cape of Good Hope, but not along the most direct route from New Zealand. He could leave Admiralty Bay and set a course to the north-west which would take him clear of the undelineated coast of New Holland, through the channel described by Dalrymple as having been taken by Torres, and into the charted sealanes of the East Indies. However, a few days sailing due west from where the *Endeavour* lay, stood an uncharted coast, a blank on the maps, needing to be filled in to complete the outline of New Holland.

Banks had noted earlier that a 'roast beef' mentality had developed among the crew; they were anxious to be on their way home to England and Cook realized he must be extremely careful in trying to gain general approval for his personal plan. An additional part of the Admiralty's instructions recommended he proceed along a route agreed upon in consultation with his officers. Cook returned to the ship on the evening of 30 March and called a meeting at which he put various propositions: they could go back east around Cape Horn in high latitudes, which would provide further opportunity to search for the great southern continent. Cook claimed he would welcome this, but the argument against it was that it would take place during the depths of the southern winter and the *Endeavour* was in no condition to face the rigours of such a voyage; the sails were already patched, the spars and rigging could not be expected to survive the battering of the Roaring Forties, provisions were low with little chance of replenishment, and the crew was not equipped mentally or physically to face bitterly cold conditions after so long in warmer seas. An alternative passage was to travel west from this point and then far south in a great circle to the Cape of Good Hope, but the same conditions would apply as for the eastward journey and going that way would mean no discoveries. The officers were persuaded by their captain that the best solution was 'to steer to the westward until we fall in with the east coast of New Holland and then to follow the direction of that coast to the northward or what other direction it may take until we arrive at its northern extremity'.

Cook got his way with what appeared a compromise on his part. It seemed to please everybody: the seamen were off home in the least exacting latitudes, the captain would be able to present the Admiralty with another original survey — perhaps to atone for not discovering Terra Australia Incognita — and Joseph Banks could collect more of his flowers in virgin territory where no European was known to have ventured before. There was little more to be said, except for the briefest journal entry by the captain for the last day of March, 1770: 'With the view at daylight in the morning we got under sail and put to sea having the advantage of a fresh gale at SE and clear weather'. By the afternoon they had cleared what they called Cape Farewell and were heading out into a heaving sea named after another great navigator, Abel Tasman. Their weatherbeaten sails would have to carry the ship for a long way yet, in fact as far as she had already travelled, because this point on the journey was as distant as they could be from England.

6
New Zealand — Australia
1770

Thus, thus I steer my bark and sail
On even keel with gentle gale.
Matthew Green (1696–1737)

METHODICAL charting of Australia was late because of its position at the very extremity of the European voyages and the pattern of winds in southern latitudes, which was almost always unfavourable to sailing ships. The first recorded sighting of any part of the continent was in 1606 by the *Duyfken*, in the Gulf of Carpentaria on the tropic north, but that was an isolated incident. Other sightings and landings occurred on the west coast. The Portuguese may have reached Australia as early as 1518 or 1522, when two expeditions were sent out to discover the so-called 'Isles of Gold', which were said to lie 100 leagues southeast of Sumatra. But as it was official policy to keep information of this nature secret, nothing was released about the success or otherwise of the expedition. These first contacts had commercial motives; seeking spices, gold and treasure to be taken back to Europe. The Portuguese also saw the saving of souls for Christ as one of their priorities; the Dutch placed less emphasis on religion, at least when exploring in distant areas. The reality appeared the same for both nations: no worthwhile goods were readily available and the native inhabitants were obviously beyond salvation. The Dutch, who became the dominant force in the East Indies, were not inherent adventurers. Their settlements were based entirely on trade and both the accidental and intentional landfalls around two-thirds of the Australian continent were entirely unproductive. The east coast remained untouched by Europeans and by the middle of the seventeenth century the Dutch had lost interest in New Holland. There was to be a gap of more than a century between the visits of Tasman and Cook.

Cook, in the words of the Roman poet Plautus, 'set his sails to suit the wind' and headed for the eastern part of Van Diemen's Land, which had been discovered and marked on the map by Tasman, although there was only inexact information about its location. By sailing due west it was hoped to arrive at the point where the Dutch explorer had departed for New Zealand. What lay ahead was mostly conjecture because, unlike much of the remainder of New Holland, the entire coast from Van Diemen's

Land to somewhere near New Guinea remained uncharted. There was occasional discussion on board about Portuguese navigators having worked along those shores more than 200 years before. There were old maps which were said to represent outlines and features of the east coast of New Holland. A French copy of a probable Portuguese original from the 1500s claimed to depict part of the coast a little north of the latitude the *Endeavour* was now negotiating, designated 'Coste des Herbaiges'. This chart had been in the possession of Edward Harley, the Earl of Oxford, and on his death in 1724 was stolen by one of his servants. Banks would eventually discover and purchase the document but at this stage he knew nothing about it, and neither did Cook. Whatever secrets had been discovered along New Holland's east coast were unlikely to have been properly recorded because instruments capable of giving accurate information were a recent development in the art of navigation. Alexander Dalrymple's *Voyages in the South Pacific Ocean* argued the existence of a clear passage at the north-eastern tip of New Holland which would allow easy access to the Dutch ports of the East Indies. Much of Dalrymple's information was wishful thinking but there were, at least, other reports that New Guinea had been circumnavigated by both the Spanish and the Dutch and the strait, said to have been negotiated first by Torres, probably existed. However, the only certainty for James Cook, 20 months out from Plymouth, was that somewhere in the direction they were heading lay an uncharted coast and he was eager to survey it.

Fair winds and fine weather carried them steadily towards their goal but then came nine days of flat calm with increasing temperatures, although they were in latitude 38 degrees south. Banks was able to engage in his favourite sport of shooting birds, including a wandering albatross, from his own little boat, while the seamen on board attended to the never-ending tasks of maintaining weakened sails and frayed rigging in preparation for the next big blow. It came on 17 April, a howling gale from the south-west that whipped up a huge broken sea causing the *Endeavour* to pitch and roll alarmingly, and take on such amounts of water fore and aft that it was hazardous for anyone but the working seamen to remain outside for fear of being swept overboard. It was discovered that the track veered too far north for finding Van Diemen's Land and Banks, ever suspicious of naval intentions, suspected the helmsmen were altering course deliberately to speed the homeward journey. 'The compass showed the hearts of our people hanging that way caused a considerable north variation', was the wry comment he wrote in his journal, not daring to discuss such a delicate matter with Cook or any of the officers. Solander was of the same opinion: 'They steered the ship rather wrong in the night, inclining to the way they wished.' The deflected course was held, and land appeared on the horizon in the early morning of the nineteenth, extending far to the south and northeast. Through a glass they could see sloping hills covered in part with trees or bushes interspersed with large tracts of sand, and three great waterspouts twisting their way between the ship and the coast. The weather became foggy and Cook was obliged to steer ENE to make certain of clearing the land ahead. The furthest visible feature to the south was named Point Hicks after Lieutenant Zachary Hicks, the first to sight this coast. Tasman's point of departure was far below the horizon and Cook had to leave unanswered the question of whether Van Diemen's land was a southward continuation of the coast off his port bow, or an island.

Next day the ship was coasting along before a brisk breeze abreast of level, wooded country with columns of smoke rising at several places to indicate human habitation. There were no discernible harbours or sheltered anchorages and the captain decided the *Endeavour* would proceed until a suitable landing place presented itself. Unable to get at the plants of this new territory, Banks spent much of his time on deck inspecting them through a telescope and his opinion veered from favourable to bad as they slowly passed by. One moment he saw extensive green fertility and the next it looked to him like the back of a lean cow, 'covered in general with long hair, but nevertheless where the scraggy hip bones have stuck out further than they ought, accidental rubs and knocks have entirely bared them of their shape or covering'. Cook was assisted by Charles Green as they made a running survey of the coast and because of few distinctive features and no useful harbours, the place names given along the way were not as dramatic or colourful as those in New Zealand. On the chart appeared the more pragmatic Point Hicks, Ram Head, Cape Howe, Mount Dromedary, Bateman Bay, Point Upright, Pigeon Hill, Long Nose, Red Point ... On 27 April, with the wind against them, they stood off the shore and launched a boat with Banks, Cook and Tupaia aboard to attempt the first landing, but a high, booming surf made it impossible to get near the beach. Three dark figures were noticed sitting naked on the shore and as soon as the yawl approached they fled into the scrub, leaving several rough canoes drawn up beside a hut, which looked to the visitors like a wigwam. From out beyond the breakers this country seemed pleasant and fertile with the trees standing free of untidy undergrowth; to the Georgian eye it was reminiscent of plantations in a gentlemen's park. Next morning the ship was edged into a broad bay that appeared to offer good shelter with an entrance wide and deep enough to allow a departure, given westerly winds. As the *Endeavour* moved through the channel there were several canoes near the shore with natives busily engaged in spearing fish. They completely ignored the arrival, as if their experience could not allow for such an event and therefore it was invisible to them. They came to anchor opposite a small settlement of six or eight dwellings. An old woman and several children looked at the ship, expressing neither surprise nor concern at its presence. A fire was lit and four canoes came back from fishing, the men landed and prepared to eat, unflustered by the *Endeavour* moored half a mile away. After taking their own dinner, landing parties were sent to the sandy, protected shore with Tupaia shouting across the water in the language of Tahiti, attempting to make contact with two natives who stood and watched, but it was obvious they could not understand what he was saying. The two blacks, decorated with broad white stripes on their bodies and foreheads, stood immobile holding spears and then threatened the visitors as they approached with exclamations that sounded like 'Warra warra wai'. Trinkets thrown across the gap of water were ignored and the naked men seemed to be daring the intruders to come ashore. Cook ordered small shot to be fired over their heads as a warning, but its only effect was to send one of them to a nearby bark hut to get an oval shield and what seemed to be a wooden sword. Then they started throwing stones at the landing party's boats, which were now reaching the shallows. The *Endeavour*'s men jumped ashore and able seaman Isaac Smith, the captain's nephew on his wife's side, had the honour of being the first among them to step onto the New Holland coast, followed by Cook, Banks and Solander. The landing party shot at the natives, who continued to throw stones, drawing blood from one of them. They finally ran off after hurling their spears at the intruders, one of which landed between Parkinson's legs, much to his and his colleagues' surprise. After this, attempts at friendly contact with the blacks were fruitless as they proved shy, refusing to accept gifts and when approached, disappearing into the vegetation as nimbly as timid deer.

The bay was broad and well protected from all quarters except the west, surrounded by low land covered with flowering shrubs. Mangroves flourished along the indented shoreline and the sandy shallows were the home of giant stingrays weighing as much as 300 pounds each. They, together with oysters and mussels, provided plenty of welcome fresh food for the crew, and, from the large middens of empty shells in the vicinity, appeared to be one of the natives' main sources of sustenance. The land animals acted as elusively as the Aborigines. Banks' greyhounds chased a small quadruped into the bushes but could not catch it, and the group also noticed the fresh dung of an

Two Aboriginal warriors at Botany Bay. After Parkinson.

arkinson del.   T. Chambers

animal that might have been a deer, as well as the cast-off skin of a large snake. Birds were in full evidence, however; the brightly coloured parrots contrasting with sombre black birds, whose glistening feathers made them look like English crows. Parrot pie tasted good after the monotony of ship's rations and even crow became palatable as a change of diet. The shores of the bay were a paradise for the naturalists and for the first time since Tierra del Fuego, fifteen months before, Banks and Solander were able to spend long periods of collecting without interruption from natives, whether friendly or hostile. It was an exhilarating time for them, although they were hard pressed for space to store so many new specimens. Several new genera were discovered and an overwhelming number of new species. Many of the them, such as spiky Red Honeysuckle, called later *Banksia serrata*, and *Isopogon anemonifolius*, or Drumsticks as it is known, with brown, woolly cones, were stranger than Solander or Banks had ever seen. Sydney Parkinson was fully employed trying to detail every plant that came aboard, an almost impossible task. There was barely time to make rapid sketches of a specimen's general proportions with additional colour notes about buds, stems, leaves and fruit before it wilted in the warm weather. Solander's classification was written later on the back of each sketch and the locality where the collection was made also noted. This botanical shorthand was the only way of dealing with an avalanche of work. Cook displayed the English colours ashore each day so that the natives might become familiar with them and he also had details of the *Endeavour*'s arrival carved in the bark of a tree. The captain surveyed the bay in his usual meticulous fashion while the men caught fish, took on fresh water and harvested the rather wiry native grass that served as hay for the vessel's sheep. Banks and Solander eventually went their own separate collecting ways because of the great variety of plants in the area. To preserve their specimens, they needed to be placed between sheets of drying paper, which, having absorbed some dampness on board, had to be returned to normal condition by being taken out to dry on the sand in the warm May sunshine. At night, moving lights were seen in different parts of the bay where natives went fishing in the same manner as had been observed in the Society Islands.

After the rigours of the New Zealand coast and the need to be on constant guard against the unpredictable Maoris, the serenity of this anchorage and the lack of interference by the Aborigines made for a pleasant stay. The food was good: parrot pie, stingray cutlets, fresh oysters, mussels and wild spinach provided excellent fare and the late autumn weather smiled on them. As preparations were made to leave and continue the voyage, Cook wondered what to call the place. The north and south sides of the bay's entrance were named after Solander and Banks because of the rich bounty of plants, but as the most spectacular feature of the area was the presence of giant stingrays, Cook noted in his journal 'Sting rays harbour'. It was colourful but not quite right, and several weeks would pass before he finally decided on another name, the one which would become the most famous of the entire voyage: Botany Bay. Just before they left, a young seaman who had contracted tuberculosis at Tierra del Fuego finally succumbed to the chronic disease. Forby Sutherland's body was carried ashore and given a decent burial in this peaceful spot, on the opposite side of the globe from his birthplace in the Orkneys.

The *Endeavour* sailed north aided by a strong southerly, passing a 'bay or harbour' where there seemed to be an extensive anchorage. Cook named it

Port Jackson — the future site of Sydney. Pressing on, the land became higher and more broken, sandy and less fertile. Sydney Parkinson, in a marathon burst of sketching, dealt with all the plants they had taken on board at Sting rays harbour and Banks was able to write in his journal of 12 May with some satisfaction, 'This evening we finished drawing the plants got in the last harbour, which had been kept fresh till this time by means of tin chests and wet cloths'. In two weeks the diligent young draughtsman had completed 94 sketch drawings. After seventeen days at sea they approached a bay which appeared rather barren with its sands fringed by low shrubs. The crew were in good humour, well-fed and thrilled with the brisk progress of their ship which coasted well in these conditions. That night, however, there occurred what Cook described as 'a very extraordinary affair' that demonstrated once again the tensions the voyage brought about and the volatility of the men's tempers when confined for so long within the heaving confines of a small ship. The captain's clerk, Richard Orton, midshipman Patrick Saunders and a group of sailors had engaged in a long drinking bout and eventually Orton reached a state of insensibility. Then, as Cook related in his journal, 'Some malicious person or persons took the advantage of his being drunk and cut off all the clothes from his back, not being satisfied with this they some time after went into his cabin and cut off part of both his ears as he lay in his bed'. The prime suspect was the able seaman James Mario Matra, who was known to behave aggressively on such occasions. He and Orton were no great friends and once before Matra had cut the clothes off the clerk's back in similar circumstances and was heard to mutter threats about doing it again. Cook was determined to discover the culprits, but there were no witnesses to the mutilation who were prepared to talk and, therefore, no direct proof. Matra, however, was held to be mainly responsible and had to accept being banished from the quarter deck, where he had been allowed as a sort of honorary midshipman. Cook, in a rare show of vindictiveness, described him as good for nothing and 'one of those gentlemen, frequently found on board King's ships, that can be very well spared'. Patrick Saunders received no words of censure from his captain; instead he was summarily demoted to able seaman because of his probable participation in the bizarre incident.

At daylight, Banks and the captain went ashore and saw a few natives who proved as coy as ever and there was no contact. Parkinson looked around finding little to interest him except a few plants, and he could do without illustrating any more of those at the moment, although he took some specimens for his own collection of natural history. He also came across the dung of some large animal that fed on grass but the creature itself was not to be seen and he concluded that most of the wildlife of this country was as unapproachable as its human inhabitants. The value of the young man as a member of the scientific team extended far beyond the duties of a mere draughtsman and Banks was delighted with his intense enthusiasm and curiosity about all aspects of the natural world. Several porpoises played around the ship as the crewmen tried without much success to net fish while Banks waded through the soft mud of a mangrove swamp in an energetic quest for something new. He was greeted by hordes of green ants after he disturbed some nests and discovered they were 'sharper than any I have found in Europe'. The leaves of the mangroves were crawling with small hairy caterpillars, a 'wrathful militia' which fell on the unsuspecting intruder with 'every hair of them stinging much as nettles do, but with a more acute 'tho' lasting smart'. The plants here were not all new species as had been the case previously; some were similar to those of Tahiti and others were known by Solander to grow in the East Indies, but Banks did find some interesting eucalypts. There were nervous pelicans keeping their distance, noisy ducks with brilliant plumage and a large black, white and brown bird — 'as large as a good turkey' — which was found to weigh 17 pounds after being shot. This bustard provided an excellent feast, 'not only good but plentiful' and also named the place for Cook's chart — Bustard Bay.

On 24 May they continued along the coast and neared the tropics with increasingly warm and humid weather. But the winds soon died and the *Endeavour* sat becalmed; the only activity while sitting and waiting for a breeze to spring up was fishing over the side with lines. There were more pelicans near the shore and somebody caught sight through a telescope of the tail of a large creature vanishing into the bush. Under way again, the tropic of Capricorn was passed and the land seemed to get more desolate as they proceeded northwards, with little to see except sandhills and rocks. The ship was gradually being drawn into an area of shoals which became obvious with increasingly erratic readings between one throw of the line and the next. It was now necessary to have a sounding boat ahead most of the time to carry out continuous depth measurements. Nobody could have alerted James Cook to the dangers of the Great Barrier Reef because there were no records of its existence. To his increasing consternation he found sea areas with as little as two-and-a-half fathoms of water, forcing regular putting-about and attempts at a different course. Whichever track was chosen there were islands all around and the utmost caution was required in navigation, resulting in a painfully slow passage after the brisk progress of the previous weeks.

On 29 May the ship passed into a bay that seemed to be the entrance of a river because of the strong tide, falling as much as 12 feet in a six-hour period. Cook thought it might be a good place to run the vessel aground to clean her bottom of weed and other marine growth but a quick survey revealed a lack of fresh water and then it was realized this rather barren place was, in fact, a channel separating islands from the mainland and not a river mouth. They stayed here for two days, allowing Banks and Solander to collect some new plants in the face of ravenous mosquitos and flesh-gripping sand burrs. Banks pursued his interest in the eucalypts, collecting a specimen of the Narrow-leaf Ironbark, or *Eucalyptus crebra*. There was great beauty in this tropical setting with little mud skippers darting like English minnows through the mangroves, flocks of brilliantly hued lorikeets wheeling overhead and a great cloud of butterflies that caught Banks' imagination: 'The eye could not be turned in any direction without seeing millions and yet every branch and twig was almost covered with those that sat still'. The naturalists wanted to spend more time ashore, but Cook saw no advantage in delaying the journey, apprehensive of the navigational challenges that lay ahead. Banks took back to the ship a little silver chrysalis and it hatched the next day into a beautiful butterfly 'of a velvet black changeable to blue, his wings both upper and under marked near the edges with many light brimstone coloured spots'. This metamorphosis might have seemed like a good omen for their future travels. It was not.

Had Cook known what faced him on this uncharted coast he would have attempted to put far out to the east in the open Pacific to escape the shoals and coral that made up the labyrinth of the reef. From a hill at the entrance to the waterway he had named Thirsty Sound, for obvious reasons, there had been as many as thirty islands in view. That could only mean a tortuous journey because of the amount of sounding needed every league of the way. On the first day of June there was wind and rain so strong that Banks found himself wondering if the anchor cable would snap. Some mornings he awoke to the disturbing sight of land in every direction and at other times there was clear sailing in an apparently open sea. Spirits should have been high on board with the European civilization of Batavia only a matter of weeks away, but there was foreboding. Tupaia showed early symptoms of scurvy with swollen gums and was ordered to take lemon extract in all his drink. Cook went about the business of navigation, assisted by senior officers and Charles Green, in his usual professional manner, showing little outward emotion or concern, although everyone realized that the captain was being tested to the limits of his experience in keeping the *Endeavour* from being grounded. The gentlemen were at supper on Sunday, 10 June when the ship passed over coral at about seven fathoms. Everyone assumed it to be the end of a line of reefs seen at sunset and they went off to their cots secure in the thought that no dangers would threaten that night. 'Scarce were we warm in our beds', Banks recalled later, 'when we were called up with the alarming news of the ship having been fast upon a rock, which she in a few moments convinced us of by beating very violently against the rocks'. The situation was serious because they had been standing off from the coast under a pleasant breeze for more than three hours and there would be little hope of getting back to the shore if the ship foundered. The contemporary English hymn-writer and poet, William Cowper, would describe their predicament perfectly:

But oars alone can ne'er prevail
To reach the distant coast;
The breath of heav'n must swell the sail,
Or all the toil is lost.

She was stuck firmly on the coral at high tide, the worst kind of obstacle because the sharp points could cut a ship's timbers like a knife through butter. Fortunately the sea was calm and the moon shone down on their plight from a cloudless sky. Cook was on deck almost immediately after the vessel struck, pulling on his trousers, ordering sails to be taken in and the boats launched with lines attached to the ship so that the *Endeavour* might be pulled off as soon as the tide rose again. There was no panic, and Banks, who had staggered sleepily up to the quarter deck, watched with admiration as the officers and sailors calmly went about their emergency routines. All the time the ship's keel continued to scrape on the jagged coral, rising and falling with the Pacific swells which, although light, were enough to cause damage. There was an ominous grinding sound from below and it was difficult to remain on two feet because of the lurching. Banks noticed timbers floating away on the moonlit sea; they were sheeting boards from the side of the ship, and then around midnight parts of the false keel also broke off.

The most urgent task was to lighten the vessel as much as possible so that there would be the maximum buoyancy. Everything heavy and disposable was hurled over the side: ballast, water casks, firewood and the six large guns on deck together with their carriages. There was little time to reflect on the seriousness of the situation as everybody was fully commited to a communal act of survival. The fact was they were far from the nearest European civilization in the East Indies, off an unknown coast where, even if the present problem was overcome, survival in the face of hostile natives and lack of supplies was by no means guaranteed. And of course the prospects for being discovered and rescued were minimal. But at least the weather remained on their side. A flat calm and the falling tide allowed the ship to sit squarely on the bed of coral at first, then, with the sea falling further, she lurched over on her side and began to take water into the hold. At first light, after what seemed an interminable night, land could just be discerned on the horizon about 20 miles distant. The ship's boats now ran out all five anchors and dropped them at different points so that the *Endeavour* could pull against them on the rising tide and, perhaps, slip off the reef. The pumps needed to be manned continuously to stem the inflow of seawater and Banks noticed that all the seamen worked around him with surprising cheerfulness and alacrity: 'No grumbling or growling was to be heard throughout the ship, no not even an oath although the ship in general was as well furnished with them as most in His Majesty's service!' As the tide rose the vessel resumed her alarming scraping against the coral and so much water rushed in that the three usable pumps could only just control the flow. Banks contributed as energetically as everybody else to the demanding exercise, but instead of sinking exhausted to the deck before the next pumping session he used the time to gather what he could carry of his personal belongings and prepared for the worst. The ship was almost afloat on the rising tide and Cook expected she would slide into deep water off the edge of the reef; the danger then would be if the pumps did not hold out, for in that case the *Endeavour* would sink. Banks considered this possibility: 'We know that our boats are not capable of carrying us all ashore, so that some, probably the most of us, must be drowned: a better fate maybe than those would have who should get ashore without arms to defend themselves against the Indians'. His darkest thought was the prospect of reaching land and then being confined there, as if in a prison, debarred from the hope of seeing England again or conversing with any but what he considered the most uncivilized savages in the world. He had also understood that as soon as a ship got into a desperate situation it was usual for the men to mutiny and plunder everything. The *Endeavour* proved an exception because of the 'steady conduct of the officers', who remained perfectly composed during the ordeal.

The moment of truth approached as the capstan and the windlass linked by cables to the anchors were manned and on a signal from Cook the men began to heave. Banks continued to pump away, refusing to accept defeat without a struggle and thinking about his plan for survival: keep the ship afloat until she could be run ashore and then build a small craft from what could be salvaged so they could reach civilization. James Mario Matra had a similar thought: 'We expected either to sink at our anchors, or be compelled to warp ourselves again upon the rocks, unless a breeze should spring up and enable us to reach the shore, where we might have so much of the wreck as would enable us to build a small bark to convey ourselves to some European settlement in the East Indies'. Between 9 and 10 p.m. of that terribly long day the tide rose four feet, the ship suddenly lurched upright again and then floated off the coral as Cook had anticipated. There followed a moment of agonizing suspense, when everyone paused, but the *Endeavour* seemed to be taking on no more water than before. Hope spread through the crew, causing even more energetic manning of the pumps. There had been 24 hours of unremitting toil and the unaccustomed hard physical labour took its toll of Joseph Banks, who had to lay down on the desk to rest. He instantly fell into a sleep of exhaustion, broken by the alarming news that the water in the hold had increased to a depth of four feet. There was a moment of panic at this revelation which, as Cook noted, 'For the first time caused fear to operate on every man on the ship'. Added to this was a breeze off the land, which meant that running the *Endeavour* ashore would be virtually impossible until the direction of the wind changed. There was a renewed and desperate effort with the three pumps, and the water gradually subsided. When there was time to make a close examination of the situation below decks it was discovered that the carpenter had made a miscalculation and the danger had appeared much more serious than it really was. The anchors were drawn in, except for one small bower which caught on the rock and had to be cut away. The cable from one stream anchor was also lost, but this was a small price to pay for the relief of floating free once again. The foretopmast was set up and they slowly got under way, heading for the coast to the northeast in the direction of two islands that were named by Cook the Hope Islands. If the unseen hole or series of breaches in the hull could be covered in some way there was every prospect they would reach the mainland. Young Jonathan Monkhouse, brother of the ship's surgeon, had previously had experience of patching a holed vessel while on a passage from Virginia to London, by a method called fothering and he suggested to Cook this should be tried on the *Endeavour*. The captain agreed and the crew got to work sewing a quantity of hair and oakum to a studding sail and spreading sheep dung over it collected from the pens on board. Cook watched the procedure and thought that horse manure would have been a better substance. Ropes were attached on either side and the foul-smelling sheet was dragged under the ship from the bow until water pressure pushed it into the broached planks and made a temporary plug. Cook was delighted with the results and was highly complimentary about young Monkhouse's initiative. The leaks could now be kept under control with one pump, and a helpful breeze sprang up as the ship was steered in search of a suitable harbour for repairs.

The first possible landing place had insufficient fresh water and it was decided not to beach the vessel at what Cook named, with feeling, Weary Bay. The pinnace explored ahead of the *Endeavour* and found a much better location at the mouth of a river, but the weather turned foul and it was four more days before the ship could get in, after scraping her damaged hull on sandbanks at the bar. During this interval of waiting to get ashore the captain had found time to have second thoughts about the affair of his clerk's severed ears and he absolved Matra of any blame, allowing him to resume access to the quarter deck, the exclusive preserve of officers and the gentlemen. There was now scurvy aboard, with Tupaia having advanced symptoms and Charles Green very poorly and not responding to any of the standard treatments, preferring to ease his pains by drinking rum. It was essential they get ashore quickly and as soon as the *Endeavour* was inside the river mouth she was beached on a steep bank, tents were set up on the sand, the sick transferred to them, and all cargo and provisions offloaded along a wooden stage constructed from the ship to dry land. Banks and

Solander immediately went off in search of plants while Cook climbed the steep hill behind the beach to look out on a view that was 'a very indifferent prospect'. The wide tidal estuary below gave way to a much narrower channel snaking its way as a river to the blue haze of the inland fringed by dark green mangroves. The *Endeavour*'s bow was pulled far enough up on the shore to reveal the pierced hull for the first time and Banks described the damage: 'In the middle was a hole large enough to have sunk a ship with twice our pumps, but here Providence had worked most visibly in our favour for it was in great measure plugged by a stone which was as big as a man's fist'. They had been both threatened and saved by the jagged spike of coral that broke off and remained lodged in the hole when the ship was freed on the tide. As well as the breach with some fothering still attached, there were deep gashes across the bottom planks, looking as if a sharp axe had been taken to them. The false keel was disintegrating and the protective sheathing to deter the teredo worm, a notorious eater of ship's timbers, was stripped away in parts. What remained of the false keel was in such poor condition that Cook remarked, 'We should be much better off if it was gone also'. He had to assume that the sternquarters were relatively undamaged, because it was impossible to bring the entire keel out of the water for a thorough inspection. As the ship was being hauled ashore, all the water in her hold had rushed to the stern, so she became dry forward but with nine feet of seawater abaft. This, as Cook expressed it with only the hint of an overtone, 'was very near depriving the world of all the knowledge which Mr Banks had endured so much labour, and so many risks to procure'. When the danger had first become evident, the botanists had transferred the entire collection of plant specimens to the bread room, deep in the after part of the ship, as the place of greatest security. Nobody thought of them when the head was brought higher than the stern, water-logging much of the collection. Banks was too dismayed to show his feelings about the incident when he realized what had happened; instead he and his team carried all the sodden plant material to shore, where, as Cook observed sympathetically, 'Most of them … were, by indefatigable care and attention, restored to a state of preservation, but some were entirely spoilt and destroyed'. Robert Taylor's forge was set up on the beach and repairs to the hull were completed in two weeks, to the industrious ringing of the armourer's assistant making nails and spikes. Furthermore, the *Endeavour* was given extra buoyancy by lashing 38 casks under her bottom, after which she was floated off the beach on a spring tide and taken across the estuary on a trial run. This proved satisfactory and she was then returned to a mooring alongside her former beached position for the re-loading of everything that had been taken ashore.

While all this activity took place around the ship, Cook was busy making surveys and Banks and Solander were striking out on increasingly distant expeditions into the hinterland, finding plenty to note and collect. Assisting in the preservation of plants were baskets made from the stalks of plantain trees, in which specimens could remain fresh for two or three days. The two men caught fleeting glimpses through the scrubby trees of the animal whose presence had been sensed several times in the south. 'He was not only like a greyhound in size and running', Banks remarked, 'but had a long tail, as long as any greyhound's; what to liken him to I could not tell, nothing certainly that I have seen resembles him'. Banks' own greyhounds could not keep up with this strange creature, which was assumed to be an entirely

new species. There were also 'wolves', or native dogs — dingoes — in the vicinity and flying foxes gliding through the trees at dusk, which were regarded by the excitable seamen as devils. Alligators lurked in the shallows near the *Endeavour*, good clams for eating came from coral reefs just offshore and many large turtles were seen swimming in the sea, although they were difficult to catch. All the food collected was divided equally among the men in a democratic gesture by Cook. 'Whatever refreshments we got that would bear a division', he explained, 'I caused to be equally divided amongst the whole company generally by weight, the meanest person in the ship had an equal share with myself or anyone on board'. He regarded this action essential for any commander on such a long voyage.

Trying to make contact with the natives had proved to be difficult because they ran away when any attempt was made to approach them. Cook and Banks discussed the situation and decided that the best course of action was to appear to ignore their presence in the hope that curiosity would bring them. These tactics were proved correct when an outrigger canoe came near the ship and the Aborigines aboard began talking in loud voices in a language that Tupaia could not understand. The officers shouted back, making signs that they would be welcome on board. They approached holding up their spears, not in a threatening way, but more to indicate readiness to defend themselves if necessary. When alongside, Cook ordered gifts of cloth, nails, beads and paper to be dropped down to them, which were met with no great enthusiasm from the natives, but when a small fish was offered, this made a positive impression. They paddled away to the shore indicating by sign language that they would fetch their companions and there followed a cordial meeting on the beach with Banks and Tupaia. Banks found it difficult to establish their true skin colour because 'they were so completely covered with dirt, which seemed to have stuck on their hides from the day of their birth'. He tried spitting on his finger and

The *Endeavour* beached at Endeavour River. Dutch handcoloured engraving after Parkinson.

rubbing the face of one native which made very little difference and he had to conclude that somewhere beneath the grime lay a chocolate-coloured skin. Cook had studied Dampier's comments about the natives on the western side of the continent in *Voyage to New Holland* (1703), most of which were uncomplimentary. This close encounter with the Aborigines of eastern New Holland was much more favourable; they had small limbs with skin of a dark brown colour, the hair black and lank. Some had their bodies painted red and others bore white streaks; Cook thought the features were far from disagreeable, with lively eyes, small, even white teeth and their voices soft and tuneful. Next morning the previous day's tentative meetings were consolidated when four Aborigines visited the ship led by a man who was distinguished by a bird bone, about six inches long, through his nose, and wearing bracelets of plaited hair on the upper arms. They were otherwise naked and when Cook gave one of them part of an old shirt he did not know what to do with it, tying it like a bandana around his head. They brought a gift of fish with them but became alarmed and left quickly when members of Banks' party began to show an interest in their canoe.

Lieutenant John Gore went on a short expedition into the bush and was able to shoot one of the strange animals that had previously eluded capture. Although a young example, it was judged to be as large as a sheep, weighing 38 pounds. Cook described its neck, head and shoulders as, 'very small in proportion to the other parts of the body'. They noted that the tail was almost as long as the rest of the animal, thick near the rump and tapering towards the end. 'The fore legs of this individual were only eight inches long, and the hind legs twenty-two'. Its progress was by 'successive leaps or hops, of a great length, in an erect posture'. It was learned from the natives that this singular creature of a 'dark mouse colour' covered with a short fur was called 'kanguroo'. John Thompson prepared it for dinner the next day and this addition to their menu was judged by both Cook and Banks to be 'the most excellent meat'.

With the main repairs and maintenance to the ship completed, the next move was to leave this coast as quickly as possible and get to Batavia where the *Endeavour* could be docked and given a complete overhaul for the long journey home. Cook and Banks climbed the hill behind the ship to see if they could discern a channel leading to the open sea and Banks noted: 'When we came there the prospect was indeed melancholy: the sea everywhere full of innumerable shoals, some above and some under water, and no prospect of any straight passage out'. They could not entertain returning the way they had come because the wind blew steadily from the south-east and the only exit seemed to be among the uncharted reefs. Banks mused: 'How soon might we again be reduced to the misfortune we had so lately escaped!' As the prospects for leaving what was now called the Endeavour River grew bleaker, a renewed sense of foreboding dampened the botanists' enthusiasm for collecting. They went about drying their plants in a desultory way while the master journeyed out of the river mouth in search of a passage between the reefs and the men continued to prepare the ship for sailing when the winds would allow. Fortunately there were plenty of fish and birds in the area and everyone ate well, especially on turtle meat. It was this delicious flesh that almost lead to a disaster with the natives. Those who had become used to the *Endeavour*'s presence were now prepared to venture on board at any time of day without fear. Ten of them arrived on one occasion, leaving their spears on shore, and were fascinated by the ship's caged birds, trying to throw them overboard, until they were restrained by the crew. Two trussed turtles then caught their fancy, lying upside-down on the deck waiting to be jointed by the cook for dinner. The Aborigines tried to steal them, thinking it was much easier to obtain their favourite food in this way than having to go to the trouble of capturing the elusive turtles at sea. Once again the crew had to restrain the natives, who became angry, left the ship in a hurry and paddled back to the shore. Banks was returning from a walk when he saw them seize their spears, gather up handfuls of tinder-dry grass, set it alight from a fire that was burning under a kettle and quickly ignite the vegetation all around the camp area. Although most of the shore settlement had been transferred back to the ship ready for sailing, one of Banks' tents, which had been erected for Tupaia when he was sick, was still standing and in the direct path of the flames. Banks shouted for assistance from the ship and the tent was dismantled just in time. Cook and his men were also able to save a fishing net and some linen hung out to dry, while keeping the natives at a distance by musket shots and drawing blood from one of them. The damage from the conflagration included a pig burned to death and the smith's forge made useless by having its wooden supports burnt away. These were minor losses, however; it was what might have happened that worried the captain. A few days earlier the entire stock of the ship's gunpowder had been drying on the sand close to the store tent with its irreplaceable contents. Had they been swept by this grass fire it would have meant the end of the *Endeavour*'s voyage and the probable loss of many men. As if to emphasize this, the blaze spread rapidly through the bush and made its way into the nearby hills so that by nightfall the sky glowed red in 'the most beautiful appearance imaginable'.

There was time to kill while contrary winds blew and made departure impossible. Banks and Solander scoured the nearby country in search of new specimens, but with little success. At one point they found some interesting nuts on the ground, which they identified as *Anacardium orientale*, although the tree from which they came was not at all obvious; they simply could not find it and after chopping down several possible sources they had to admit defeat and return exhausted to the ship. At the end of July John Gore bagged another kangaroo, weighing 84 pounds and far from fully-grown. After preparing the animal for dinner it proved a disappointment having an unpleasant taste, not at all like their previous excellent meal. On an expedition up the river with Gore, Banks had described this country as, 'generally low, thickly covered with long grass, and seemed to promise great fertility, were the people to plant and improve it'. His general opinion of what he had seen along 2,000 miles of coastline, however, was less favourable: 'Upon the whole New Holland, though in every respect the most barren country I have seen, is not so bad that between the productions of sea and land a company of people who should have the misfortune of being shipwrecked upon it might support themselves, even by the resources we have seen'. This was the first hint that the botanist was thinking about settlement. He was not to know then that in less than 20 years Botany Bay would become a name to strike terror into the hearts of English convicts, a synonym for exile and punishment.

The strong winds continued until the end of the month when they suddenly fell calm, followed by a light land breeze. Cook sent a boat out to the bar to check the depth of water while the anchor was raised and the *Endeavour* made ready to sail. Spirits were buoyant again with a healthy

Capparis lucida Watercolour drawing by Frederick Polydore Nodder of an Australian specimen collected at Booby Island.

crew anticipating the delights of civilization, although enthusiasms were tempered by the knowledge of treacherous coral reefs just outside this safe harbour. Banks suspected that some of the officers and men were stalling the departure for fear of the unknown, but once again his thoughts had to remain private; spoken openly they might have sounded like accusations of mutiny. The boat returned to the ship and the officer reported only 13 feet at the bar; it was six inches less than the vessel's clearance. They were obliged to come to, strike the sails and abandon any hope of an immediate departure. Several frustrating days would pass before they could leave and in the waiting time there was little to do but catch fish and carry out routine shipboard maintenance. The gentlemen continued their botanical classification and Parkinson tried to keep ahead of his quota of drawings. The delay also allowed time for a routine examination of the pumps by carpenter John Satterley and it revealed they were all in a bad state of decay because, apparently, the wood had never been properly cured. One pump that had been inoperative for some time was so rotten when hoisted out that it fell to pieces on the deck, causing Cook to comment with resignation, 'Our chief trust is now in the soundness of our vessel, which happily does not admit more than one inch of water in an hour'.

Another unsuccessful attempt was made to depart from Endeavour River and it was not until the early morning of 4 August that escape was realized. Sail was raised to catch a light air coming from the land as the ship slowly edged her way out of the estuary and, preceeded by soundings all the way, moved to an anchorage over a sandy bottom in 15 fathoms of water. Cook was not prepared to go any further until he knew if it was safe to run among the shoals, and that could only be assessed at low water by viewing

Barringtonia calyptrata
Eighteenth-century proof engraving by Gerald Sibelius from an Australian specimen collected at Endeavour River.

the scene from the masthead. Then an important decision had to be made: whether to attempt to beat back to the south along the route they had already come and strike out for the open sea, or seek an onward passage east or north, both of which appeared equally difficult and dangerous. Cook was in favour of a north-east track, past a point in the distance he named Cape Bedford, in the direction of what looked like two large islands. He thought this would offer the least interrupted journey, but he was not to know that such a passage would become an even greater nightmare of reefs and shoals, providing the worst experience of the voyage so far. There was a strong gale on the morning of 6 August so that instead of weighing anchor, more cable had to be paid out to hold the ship's position and the top gallant yards were taken in. Cook and his senior officers climbed to the swaying masthead at low tide and all they could see to the east was a continuous line of surf breaking on the reef. The captain had to admit that he was now at a loss to know which way to steer through this labyrinth, and that was the name he entered on his chart of the general area. The master, Robert Molyneux, believed they should try to beat back against the prevailing winds but Cook saw that as 'an endless labour'. Deliberations went on all day without resolution while the gale increased, forcing the striking of the top masts and the need to put out a larger anchor to avert further drifting. A decision had to be made about their direction and it was finally agreed to proceed along the coast, past Cape Bedford, as originally suggested by Cook. The wisdom of this seemed to be confirmed when a clear passage was revealed by the sounding boat and it looked as if progress was possible, but it was not long before the passage became blocked again. The distant land which had raised everybody's hopes was awarded the name of Cape Flattery.

Cook thought he should visit one of the high islands ahead to get a better view of the surrounding seas from its top and he and Banks set out in the pinnace while the master was sent in the yawl to make soundings between their present position and the mainland. On their way they chased several green turtles without being able to catch one and then soon after midday reached the island. Cook immediately set out to climb to the top while Banks, 'whose fortitude and curiosity made him a party in every expedition', according to the captain, started collecting plants. The summit was approached with a mixture of hope and fear; what could be seen was enough to boost both emotions. There were shoals all around, and in the misty distance, two or three leagues away, was a long reef of coral rocks on which the surf pounded. Apparent gaps were seen in the reef through which the tide was surging. The strong wind was causing a sea mist, which reduced Cook's visibility and since he needed to be absolutely sure of what he was seeing he decided to wait until the next day because it was now getting close to sunset. Banks was happy to stay on the island overnight so that the captain could make another observation early the following morning. The small party camped out in the open on the sand of Lizard Island, named for the only living creature they had seen there — apart from the ever-present and irritating sandflies. The island was small and barren although there were the frames of some native huts and vast pits of discarded shells. Banks noted a few plant specimens, including the freshwater mangrove, *Barringtonia gracilis*. They settled down for the night under the shelter of a low bush and Cook rose at 3 a.m. to set out on his climb up the 1,100 foot hill again, but on reaching the top found the weather even hazier than the day before. The pinnace returned at noon having been over to the mainland and out to the reef. Robert Molyneux was able to report deep water all the way, although the channels to the open sea appeared to be rather hazardous for a vessel of the *Endeavour*'s size. Cook was well aware that getting through a narrow gap into the ocean was going to be dangerous because of the difficulty of steering the ship in the face of the prevailing winds and the possibility of striking the coral again. 'An accident of this kind', he admitted, 'or any other that might happen to the ship would infallibly lose our passage to the East Indies this season and might prove the ruin of the voyage'. On returning to the *Endeavour* he discussed the situation with a committee of the officers, emphasizing that only three months' provisions were left and it was imperative to take on supplies and get repairs as soon as possible. The unanimous decision was to head out to sea. At daybreak on the morning of 13 August the anchor was weighed and the vessel approached a gap in the reef.

There was a steady gale at south-east and by the early afternoon they were getting close. Cook tacked the unwieldy *Endeavour* while the master checked and rechecked until he was satisfied that conditions were at their best for the attempt. He signalled them to proceed and they passed through without incident. 'As soon as the ship was well without it', Banks wrote in his journal, 'we had no ground with one hundred fathoms of line so became in an instant quite easy, being once more in the main ocean and consequently freed from all our fears and shoals'. Cook observed that this single act of escape generated a completely different atmosphere on board: 'Our change of situation was now visible in every countenance, for it was most sensibly felt in every breast; we had been less than three months entangled among shoals and rock, that every moment threatened us with destruction: frequently passing our nights at anchor within hearing of the surge that broke over them; sometimes driving before them while our anchors were out, and knowing that if by any accident, to which an almost continuous tempest exposed us, they should not hold, we must in a few minutes inevitably perish'. Cook committed many such thoughts to his journal, no doubt relieving himself of some of the strain he had been under all these weeks.

Next morning they were out of sight of land for the first time in nearly three months and spirits were high once again; perhaps too much so for their own good. Cook was not going to stand out to sea for long because he was determined to settle the question of a clear passage between the northernmost part of this country and the island of New Guinea and he did not want to overshoot the strait. What he called 'a well-grown sea' came in from the south-east and broke on the reefs, which again were visible in the far distance. The wind moved around and this meant it could drive the ship closer to the breakers if it persisted in blowing onshore. All sail was set and the *Endeavour* continued her northerly course with everyone knowing that if anything went wrong and they were forced any nearer to the reef, 'clearing it would be doubtful', in the words of their captain. He wondered if they were going too far to the north for Torres Strait and at midnight ordered the ship to be tacked and stood back to the south, but then the wind quickly fell to a calm and the pitch black night was spent in anxious sounding without touching bottom with 140 fathoms of line. By 4 a.m., however, there was the ominous and growing sound of surf breaking on a reef and by first light it materialized as a foaming white barrier barely a mile away. Banks had noticed the previous day, when looking through his glass, 'The large waves of the vast ocean meeting with so sudden a resistance make

WORMIA ALATA Br.

F. P. Nodder pinxit 1778.

Dillenia alata Eighteenth-century proof engraving by Gerald Sibelius from an Australian specimen collected at Endeavour River.

here a most horrible surf, breaking mountain high'. The situation was now getting more desperate by the minute. With no wind to check a drift to possible oblivion the only hope was whether the *Endeavour*'s boats would be able to tow the ship away from danger. The pinnace was under repair and could not be launched, but the yawl and the longboat were put into the sea and after much exertion at the oars the sailors managed to pull the ship's head around to face the north, an action which, Cook remarked bluntly, 'If it could not prevent our destruction, might at least delay it'. By the time this manoeuvre was completed the vessel was barely 100 yards from the wall of the reef, although the sea beneath them was still unfathomable. On

102

the quarter deck Charles Green, apparently oblivious of the impending doom, continued to take sextant readings assisted by gunner Forwood. The gap narrowed and they found themselves only the width of a trough between waves away from destruction. Banks thought it was the end: 'A speedy death was all we had to hope for and that from the vastness of the breakers which must quickly dash the ship all to pieces was scarce to be doubted'.

But fate favoured them once again. A light land breeze sprang up, a mere catspaw that hardly ruffled the surface and would have been ignored in normal circumstances, but this 'friendly little breeze' was enough to turn the scale and give the ship a slight momentum away from the reef at an oblique angle. It was a flighty little zephyr, however, which quickly dropped again before recovering for a few minutes. A small opening in the reef was noticed about a quarter of a mile away and one of the mates was sent to make an assessment. He reported its width as barely the length of the *Endeavour*, but there was smooth water on the other side. Cook decided not to take the risk, as the *Endeavour* was managing to stay off the reef, though in a perilously close situation. Another gap came into view and Lieutenant Hicks was sent to investigate while the vessel struggled to hold her position against a strong flood tide. He returned to report that the opening was passable. Cook had to make a move. He headed the ship towards the gap and, caught up in the swirling tide, she was carried through by the surge like a piece of wood bobbing in a mill race. As soon as they reached the smooth water inside, the anchor was dropped in 19 fathoms on a bottom of coral and shells. Richard Pickersgil, the master's mate, one of several men who were keeping journals of the voyage, noted with relief that what they had just experienced was 'the narrowest escape we ever had and had it not been for the immediate help of Providence we must inevitably have perished.' The gap was noted on the chart as Providential Channel, though they now found themselves back in the same situation, back inside the reef, that Cook had tried so hard to escape only two days before. 'Such is the vicissitude of life', was the resigned comment from the captain, but when he got around to writing an account of the day's happenings he engaged in some more revealing soul-searching. Cook agonized over problems inevitably faced by an explorer. If he did not fully examine newly-found coasts he was open to the charge of being timorous. On the other hand, if he boldy faced all the dangers and obstacles and then found nothing new he would also be accused of temerity. He admitted to himself that recent travels through the coral should not perhaps have been carried out with a single vessel for the reasons of survival in the event of shipwreck.

On the morning of 18 August they set sail on the next stage of the journey through the shoals and low-lying islands, all of which were surveyed as the ship slowly passed. The *Endeavour* rode at anchor each night because it was unsafe to be under way during darkness and any further dangers must be avoided at all cost. Three days later there were no more islands in the ship's path and Cook ordered all sail to be carried as she headed for the northermost tip of land on the horizon. A few islands appeared again in the far distance but the mainland on their port bow ended by dropping away to the west. The pinnace was sent to the shore, where a few natives stood and watched, but, true to previous form, they vanished as soon as the boat's company landed. The captain, accompanied by Banks and Solander climbed the small rise that passed for a hill in these low-lying parts and were able to confirm that this was the most northern part of the coast along which they had been travelling for the past four months. Cook was confident that no Europeans had been here in recent years and hoisted the English colours. Previously he had taken possession of several parts of the coast but he now thought it appropriate to annex the whole eastern seabord in the name of His Majesty King George III. The jack fluttered on the portable flagpole in the warm evening air as the marines in attendance fired a volley which was answered like an echo by others from the ship and a hearty round of three cheers from the crew up in the main shrouds and on the deck. Banks and Solander collected a few plants on their way back to the *Endeavour* as daylight quickly turned into the multicoloured splendour of a tropical sunset. Tomorrow they would sail away from New Holland. The East Indies and their first contact with European civilization for nearly two years beckoned.

Four exciting and dangerous months on this coast had produced thousands of plant specimens together with working sketches for about 412 of them. They would be developed later into finished watercolours and from these, some 340 copper plate engravings were to be prepared, forming the largest section by far of the *Florilegium.*

7
Australia — Java
1770

God moves in a mysterious way
His wonders to perform;
He plants his footsteps in the sea,
And rides upon the storm.

William Cowper (1731–1800)

THEY were approaching charted seas. Some of the *Endeavour*'s men had been in the vicinity before with Captain Wallis on the *Dolphin*, and sailors from the European seafaring nations knew these waters to and from the hub of trading activity, the Dutch port of Batavia in Central Java. With relatively clear sailing ahead there was now time for both Cook and Banks to attend to their journals and spend many hours writing impressions of the country they had just left. The captain made little attempt to review the natural history of the coast: 'Nor will this be of any loss, since not only plants but everything that can be of use to the Learned World will be very accurately described by Mr Banks and Dr Solander'. Instead, he gave a general analysis of the land and its potential for producing crops, given the right seeds, and thought that cattle could thrive in certain localities. He expressed surprise that the eastern seaboard was not as arid as the western parts described by William Dampier and the Dutch navigators. What was unexpected in his impressions was his reference to the Aborigines in a style that was out of character for such a practical man, more concerned with running a ship than thinking about aspects of anthropology. 'From what I have seen of the natives of New Holland', Cook wrote, 'they may appear to some to be the most wretched people on earth, but in reality they are far more happy than we Europeans, being wholly unacquainted with the superfluous but necessary conveniences so much sought after in Europe, they are happy in not knowing the use of them'. This Rousseau-like reasoning differs greatly from the pragmatic voice of the navigator discussing his surveys, sounding more like crumbs from a conversation with the gentlemen in the great cabin. Banks thought that the Aborigines, whose life was 'but one degree removed from brutes', had a good knowledge of plants because they had names for most of them. He noted very little about the flora of New Holland in his journal, leaving the scientific descriptions and classifications to Solander. Of plants in general he wrote, 'The country afforded a far larger variety than its barren appearance seemed to promise'.

105

But he was not averse to analyzing the Aborigines' station in life and what might be learned from it. 'From them', he wrote, 'appear how small are the real wants of human nature, which we Europeans have increased to an excess which would certainly appear incredible to these people could they be told it'. The most vivid description of the natives came from observations by James Mario Matra, who continued his clandestine journal with details of the inhabitants of Endeavour River. He saw them as 'very low of stature, commonly not more than five feet in height, small and slender in shape, but very active. Many of them had flat noses, thick lips and bandy legs. They were ignorant, poor and destitute not only of the conveniences, but also of the necessities of life. They were strangers to bread and everything that can be considered a substitute for it; nor would they eat of it when offered. They were naked and slovenly, subsisting mostly on fish, which they roasted on wooden spits stuck into the earth before a fire. We saw none of their women; but the men each had a hole through the septum nasi, or division of the nostrils, in which a bone five or six inches in length was inserted and worn as an ornament; and however ludicrous it might appear it is but just to observe, that many of our European ornaments have no more relation to natural fitness or utility, than this inexpensive one which the poor ignorant New Hollanders have invented. Besides the bones in their noses, they wear others of equal length in their ears; which though not so brilliant as the ornaments that depend from the ears of the fair sex in civilised countries, may be as useful and proper'.

Cook needed to reach Batavia with all speed, having to forgo the satisfaction of charting the geographical relationship of New Holland and New Guinea. The passage was going to be difficult enough against contrary winds without delaying their progress with too many stops on the way. It was not until 3 September, nearly a fortnight after leaving New Holland, that a landing was made on the coast of New Guinea, when a well-armed shore party of 12 was sent in search of fresh food and water. Cook, Banks and Solander stepped onto the beach and were confronted by natives armed with darts and sharpened reeds. A volley of small shot was despatched as a warning but it had no effect and a second round had to be fired before they dispersed. There were coconuts to be had on the ground, a few plantains and some breadfruit, but little else, although the soil appeared to be very fertile. Banks and Solander looked around for new plants although they could find little they did not already know. There was obviously no reason to linger here: 'As soon as ever the boat was hoisted in', Banks reported, 'we made sail and steered away from this land to the no small satisfaction of, I believe, three-fourths of our company'. According to him, the sick soon became well and the melancholy gay at the prospect of going home. Banks amused himself by observing the attitudes aboard ship now they were headed for Batavia. He was able to diagnose a singular malady 'which the physicians have gone so far as to esteem a disease under the name of Nostalgia; indeed I can find hardly anybody in the ship clear of its effects but the captain, Dr Solander and myself'. He reasoned that they had managed to escape because of 'constant employment for our minds which I believe to be the best, if not the only remedy for it'. The strong wind continued to affect the ship's speed and some of the men complained that they must be running into the westerly monsoon, which would prevent them getting to Batavia. After a couple of days, however, fair breezes assisted the *Endeavour* along again and the inveterate grumblers had to search, according to Banks, 'for some new occasion of sorrow'.

It was obvious that food was becoming scarce as the men began to eye the dolphins, barracuda and sharks around the vessel as potential meals; when two sharks were caught there was no hesitation by the crew, officers or gentlemen in eating them, the man-eating reputation conveniently forgotten. By 16 September they were off the south of Timor, carried along by brisk trade winds and that evening were treated to a remarkable display of the Aurora Australis in the southern sky: a great red glow with pulsating rays of bright light shooting up from the horizon. Next day came the first tentative contact with European civilization when the ship approached the island of Savu. It appeared at first glance to be rather unpromising with bare, brown hills, but closer examination revealed the welcoming sight of cattle grazing and Cook decided to send a boat to investigate with John Gore in command. He returned with the information that a bay existed around the next point where provisions could be bought. By evening they had arrived there, finding a village on the hillside with the Dutch flag flying. The following morning the shore party was informed by the local headman that supplies were indeed available, but only if the Dutch representative agreed to release them, so Cook had to begin delicate negotiations. The Dutchman and the headman were invited to a dinner of mutton on board the *Endeavour*, during which the headman became curious about English sheep. He was presented with a gift of the last live one on board. The greyhounds also fascinated him and Banks, in a wave of generosity, allowed him to keep the male member of his pair. The Dutch representative was, in fact, a native of Saxony, one Johann Christopher Lange, who hinted that he would like a spying glass. One was immediately presented to him. The two visitors got quite drunk and, on leaving, the headman showed interest in the armed marines, asking to see them in action. Once again his wish was granted and three rounds of shot were fired as he staggered off the boat, impressed with the speed at which the men could cock their guns.

A trading agreement had been made over the port and brandy, but the following day nothing was forthcoming. More frustrating negotiations were required together with the payment of bribes before the promised fowls, bullocks, goats, hogs, fruit and eggs were delivered. It was a hint of the Dutch East India Company's power in this region and an example of their monopolistic practices. Moreover, half the eggs proved to be rotten. In all innocence, Parkinson wanted to learn more about the island for an entry in his journal and asked about local products and, particularly, what spices were available. This raised the suspicions of Lange about commercial spying and he ordered that none of the visitors should be allowed to go anywhere in his area of jurisdiction without a strong guard. Sydney Parkinson had to confine his studies to noting a vocabulary of the native language before the ship left for Batavia, now little more than a week away. As they were travelling along the Sunda Strait, Cook followed Admiralty instructions and collected the log books and journals of his officers and men, issuing orders that no information should be given to the Dutch authorities about any aspect of the *Endeavour*'s journey. Those who had been keeping accounts of the voyage included the second in command, Zachary Hicks, midshipman John Bootie, gunner Stephen Forwood and the master's mates Richard Pickersgill, Charles Clerke and Francis Wilkinson. All of these, together with three anonymous journals, were very repetitious because the ship's log was readily available for coyping and most of the material came from that source. James Mario Matra, however, was able to secrete away his extensive accounts while Banks and his party retained the right to continue their own writings.

First news of the civilized world came from an East Indiaman encountered on 2 October. The pinnace was hoisted out and Hicks went on board to learn what had been happening in Europe while they were away. He was told that the English were rioting against King George, American colonists had refused to pay taxes and war was imminent, Poland was in a state of unrest and the Russians had attacked the Turks by sea and land. It was a gloomy summary but it did little to dampen the optimism of the homeward-bound sailors. On 10 October the *Endeavour* anchored in the Batavia Roads among 16 large cargo ships, three of which were from England, and Banks could not help noticing that the men aboard one of the British East Indiamen were 'almost as spectres, no good omen of the healthiness of the country we were arrived at'. By contrast, the *Endeavour*'s men were mostly in excellent physical conditon. Lieutenant Hicks was sent ashore to inform the governor-general of their arrival together with an apology for not giving an appropriate salute on entering the port because of a lack of guns. It could not be revealed, without giving away secrets of their voyage, that six of them lay submerged on a coral reef off the northern coast of New Holland. Hicks returned with a message that the captain was welcome and his requests for repairs and provisions would be considered at a council meeting the next day. On board the returning pinnace were pineapples, plantains and water melons, all most acceptable to sailors who had not tasted fresh fruit in such abundance since leaving Tahiti. The officers and gentlemen were also delighted with the bundle of London newspapers that arrived as well. The formal application for assistance was granted, but the cost of repairs and services was so high that Cook had to apply for a loan from the council to pay for them, which delayed the start of work until the second half of October.

Around the year 1500 the site of Batavaia, now called Jakarta, was a Hindu settlement with a flourishing trade in pepper, rice, oxen, pigs and fruit, goods which were shipped to Malacca, the leading market place for Asian and European trade. Then the Europeans came in a conquering mood, with the Portuguese taking Malacca and forming an alliance with the Hindu kingdom on Java, but their influence was short-lived and they were overthrown by Islamic invaders. At the end of the century, in 1596, the first Dutch expedition to the East Indies arrived, taking Malacca, and then looking for a base in Java. Twenty years later they built a warehouse and a fort at the river's mouth, naming the settlement Kasteel Batavia. There were sporadic local revolts against their presence but the Dutch were strong enough to quell any opposition and set about building an Amsterdam of the south, called Batavia. It was laid out to a geometric plan of streets and canals, whose water was diverted from the river. As trade in rare spices and other products grew, the town developed into an elegant colonial city attracting traders and workers from all over the world. For the remainder of the century Batavia was proudly called 'Queen of the Eastern Seas', until a violent earthquake severely blocked the canal system, breeding disease, attracting millions of mosquitos and bringing the inevitable malaria. By the late eighteenth century, when the *Endeavour* arrived, Batavia was still a flourishing centre of commerce, sending vast riches home to Holland and trading with the rest of the civilized world, but it had become the most

Banks' Florilegium Plate 54 Australia *Bossiaea Heterophylla*

Banks' Florilegium Plate 84 Australia *Castanospermum australe*

Banks' Florilegium Plate 89 Australia *Acacia legnota*

Banks' Florilegium Plate 103 Australia *Bruguiera conjugata*

Banks' Florilegium Plate 223 Australia *Ipomoea indica*

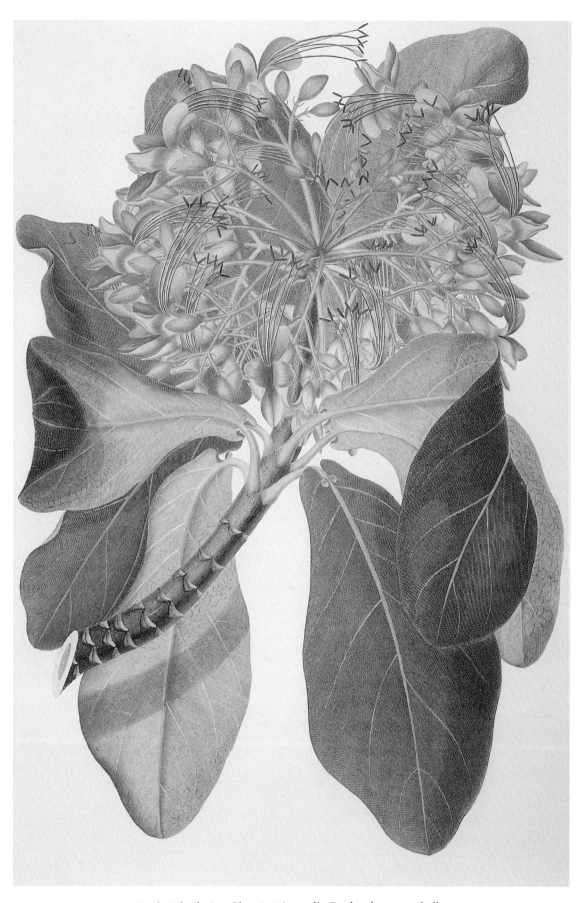

Banks' Florilegium Plate 244 Australia *Deplanchea tetraphylla*

Banks' Florilegium Plate 285 Australia *Banksia serrata*

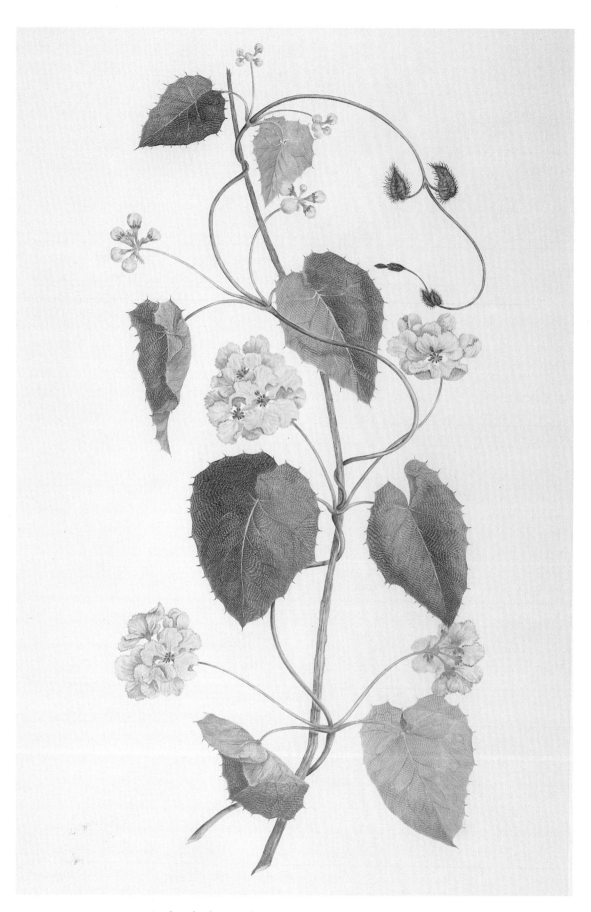

Banks' Florilegium Plate 341 Brazil *Stigmaphyllon ciliatum*

Banks' Florilegium Plate 355 Brazil *Bougainvillea spectabilis*

Banks' Florilegium Plate 371 Java *Cerbera manghas*

Banks' Florilegium Plate 395 Madeira *Heberdenia bahamensis*

Banks' Florilegium Plate 441 New Zealand *Metrosideros albiflora*

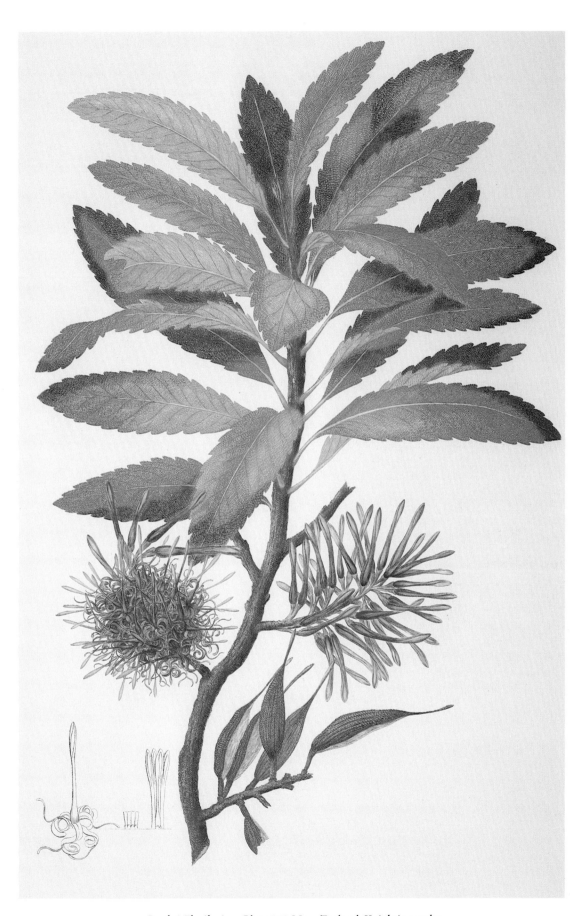

Banks' Florilegium Plate 540 New Zealand *Knightia excelsa*

Banks' Florilegium Plate 591 Society Islands *Thespia populnea*

Banks' Florilegium Plate 671 Society Islands *Cordyline fruticosa*

Banks' Florilegium Plate 726 Tierra del Fuego *Nanodea muscosa*

dangerous port on the globe for sailors and foreign residents alike. Death in the form of every known tropical ailment, but particularly malaria and dysentery, lurked in this steamy, colourful setting with tens of thousands of victims from disease each year regarded as a normal toll.

Before the ship could be handed over to the Dutch for repairs Cook needed to know her structural condition and called for a full report from carpenter Satterley. Little could be assessed of the hull without taking her out of the water but it was obvious that the main keel was holed in several places and the false keel torn away beyond midships. One pump was useless, with the others decayed to within one-and-a-half inches of their bore. Satterley's summing up of 'very leaky' was accurate. Above the waterline, masts, yards, hull and the ship's boats were, of course, able to be inspected thoroughly and were judged to be in reasonable condition. It was necessary for the ship to be hove-down and her bottom repaired before it was safe enough for the long and demanding journey still ahead. Soon after their arrival, a fleet of Dutch merchant ships was about to set off for Europe and two important sealed packets from the *Endeavour* were entrusted to the commander, Captain Kelger, aboard the *Kronenburg*. One was destined for the Admiralty in London and included a letter to Secretary Stephens taking the form of a precis in about 600 words of Cook's journal from the time of leaving Rio de Janeiro to arrival in Batavia. Apart from the geographical details of the voyage, the captain praised the excellent work of astronomer Charles Green and noted the significant natural history discoveries of Banks and Solander. There was also appreciation of his own officers and crew: 'They have gone through the fatigues and dangers of the whole voyage with that cheerfulness and alertness that will always do honour to British seamen'. A copy of Cook's journal in Richard Orton's hand was included in the packet, together with charts of the South Seas, New Zealand and the east coast of New Holland. Cook's letter added that he had not lost one man from sickness during the voyage—which meant from scurvy—and he apologized for failing to discover 'the so much talked of Southern Continent'. He claimed that much more survey work would have been done, but for running ashore: 'as it is I presume the voyage will be found as complete as any before made to the South Seas'. A second packet went to the Royal Society simply announcing the successful completion of the mission and containing some of Green's astronomical observations from Tahiti.

The ship had been moved over to Kuyper Island in the bay where her stores and all moveable objects were taken ashore and placed in warehouses in preparation for repairs at the nearby island dockyard of Onrust. Banks and his retinue went ashore and took lodgings in a hotel, which proved to be less congenial than life aboard ship, with appalling food in the midst of apparent plenty. 'Our dinners and suppers', he complained, 'consisted of one course each, the one of fifteen, the other of thirteen dishes, of which when you came to examine, seldom less than nine or ten were of bad poultry, roasted, boiled, fried, stewed, etc., and so little conscience had they in serving up dishes over and over again that I have seen the same identical roasted duck appear upon the table three times as roasted duck before he found his way into the fricassee from whence he was again to pass into forcemeat'. It was a regulation for foreigners to stay at government-controlled hotels but Banks was able to circumvent this by renting the house next door and employing servants so that his lifestyle was considerably improved.

After settling in he called for Tupaia and the boy Taieto to join him in town
and they were immediately captivated by the vivid sights of Batavia as
Banks showed them around in one of his two hired carriages. The Tahitian
priest had been sick with a stomach infection for a long time and rejected
any help from the ship's surgeon, but now a change of air and the ability to
be out and about in this exciting environment brought an immediate improvement to his health. Banks observed: 'On his arrival his spirits, which
had been very low, were instantly raised by the sights which he saw, and his
boy Taieto, who had always been perfectly well, was almost ready to run
mad'. They were amazed to see horses drawing carriages and large, permanent buildings in wide streets. Many of these things had been described to
Tupaia by Charles Green, who spent a considerable time teaching him
English, but he had been unable to relate them to his own experiences in
Tahiti, 'So he looked upon them all with more than wonder, almost mad
with the numberless novelties which diverted his attention. From one to the
other he danced about the streets, examining everything to the best of his
abilities'. One day while Banks and Tupaia were out walking a man rushed
up to them and asked if the Tahitian had not been in Batavia before; Banks
confirmed that he had not and was curious to know why such a question
should be asked. Apparently Bougainville had been here a year-and-a-
half before with two French ships, *La Boudeuse* and *L'Etoile* and a native
who looked just like Tupaia was travelling with them. That was how the
Endeavour's officers learned of Bougainville's expedition in the Pacific and
it explained the presence of the mysterious vessel at Matavai Bay, which the
natives had said was flying a Spanish flag. After hearing this story and
learning that a botanist, Philibert Commerson, and his draughtsman pre-
ceded them, Banks had to admit, 'They probably have done some part of
our work for us.' But this was not so. The botanical records of Bougainville
in the South Seas had produced only a superficial record of the region's
flora and the naming by Commerson, of one of its most beautiful species,
Bougainvillea, after his commander.

There was plenty to experience for all members of Banks' party in this
bustling centre of tropical commerce. Sydney Parkinson was as energetic as
his master in getting about the city and his personal journal became an
excellent guide-book to late-eighteenth century Batavia. He noted it was
walled-round, with canals cut through, supplied by a river divided into
several streams. The roads were, he thought, as many and as good as those
in England, stretching out into the country and bordered by vividly green
durium, breadfruit and coconut trees, with many spacious villas nestling
between them. Parkinson saw the entire countryside as a garden, divided
into different plantations by hedgerows and canals with an order that had
much more to do with Dutch colonists than the native inhabitants. With no
intimation of his own mortality or that of his colleagues, he noted that the
useful and picturesque waterways were supposed to be prejudicial to the
health of the townspeople because they became sluggish during the dry
season and then putrid, filling the air with noxious fumes. The thick fringes
of trees beside the canals looked attractive enough but they were said to
contribute to the problem by preventing the vapours generated by the
stagnation from being dispersed in the generally languid breezes. He learned
that fever often raged among the Batavians and that the slaves brought from
India were also affected. The fever, which often carried off victims within a
few days, was malaria and, together with the other great killer of foreigners

Apelocissus arachnoidea Watercolour drawing by James Miller of a vine from Java.

known as the flux, made this place one of the unhealthiest in the world for Europeans. The locals regarded the flux, or dysentery, as a relatively trivial matter, but few visitors escaped its agonizing effects. Parkinson, ever the believer in his own brand of moral justice, thought this was simply because most Europeans ignored strict rules of cleanliness or did not care enough about what they ate.

It had been two years since the *Endeavour* was in a large centre of population, but Rio de Janeiro had been unreasonably restricted to the passengers and crew and so Batavia was the first port of call since Funchal that allowed a taste of European living, in spite of its suffocatingly hot climate. More notice might have been taken of those pale, gaunt faces that had greeted them on arrival, but the *Endeavour*'s men began their stay in

111

rosy health and nobody thought of the possibility of becoming sick. Banks and his party continued their excursions. They noted the gates in the walls with drawbridges over the canals that were drawn up for security each night, and the fine public buildings, including the town hall, castle and several churches all built in the Dutch style. The northern architecture had transplanted successfully from the soft light of Europe to the harsh brilliance of the tropics. Batavia's bustling marketplace was neatly intersected by rows of stalls piled high with luscious fruits, fresh vegetables, poultry, pork and dried fish; a separate market specialized in shellfish and fresh meat. The city was the seat of Dutch administration in the East Indies and they had made it one of the most flourishing centres in this part of the world. Indeed, there was perhaps no such place anywhere, according to Sydney Parkinson, that contained a greater variety of people. Many of the local people retained a traditional style of dress and were allowed to live in their own fashion. The majority of workers in the white population were not Dutch, but Germans, Danes, Swedes and Hungarians; the main foreign merchants, apart from those of the Dutch East India Company, were English, French and Italian, displaying their affluence with carriages and slaves and a flamboyant style of dressing that flaunted silk, velvet and lace with finely-dressed wigs for appearances in public. However, they and their wives and families displayed pale, even sickly, complexions, all the more noticeable in the tropical setting, surrounded by so many dark-skinned people.

It took less than a fortnight for malaria to start showing its effects on the *Endeavour*'s company. Banks wrote: 'We began sensibly to feel the ill effects of the unwholesome climate we were in: our appetites and spirits were gone, but none were yet really sick except for poor Tupaia and Taieto, both of which grew worse and worse daily, so that I once more began to despair of poor Tupaia's life'. The Tahitian was so desperate that he pleaded to be returned to the ship where he thought he might be able to breathe, away from the stifling house in town, which he assumed was the reason for his condition. Banks took him and Taieto to Kuyper in the Batavia Roads where the *Endeavour*'s stores and equipment were being held, pitched a tent on the shore to catch the land and sea breezes and looked after them. After returning to his house in town, Banks himself almost immediately came down with malaria, 'as to deprive me entirely of my senses and leave me so weak as scarcely to be able to crawl down stairs'. Out on Kuyper Island most of the ship's men were also laid low in tents. Cook had wanted to carry out repairs with his own crew, as they were perfectly qualified to do them, but he was now forced to admit the impossibility of this, for even if the authorities allowed it, he could hardly muster 20. The ship's keel was found to be in a worse condition that anyone imagined. An area of six feet in length and two-and-a-half planks wide, which could not previously be examined properly and remained covered by water at Endeavour River, had been within one-eighth of an inch of being cut through by the coral. The Dutch supervisors could not understand how the *Endeavour* had remained afloat and Cook was relieved that he and his officers had been blissfully ignorant of the real state of the keel for all those hundreds of miles. In mid-November, with the main repairs completed, the *Endeavour* was returned to Kuyper Island for three weeks of re-rigging and taking on of stores and equipment.

William Monkhouse became the first of the Batavia malaria casualties and the position of surgeon on the ship was taken over by his assistant, William Perry. Solander attended Monkhouse's funeral while Banks was too weak to leave his bed. The young Tahitian Taieto was the next to die, of a combination of malaria and dysentery. He had taken all the medicines offered to alleviate the condition, to no effect. Tupaia continued his refusal to accept anything, and grief at the loss of his boy hastened his own demise a few days later. They were buried together on the nearby island of Edam beside the rope factories and the prison for convicted European felons, an unhappy conclusion to their strange journey. Cook had never really been at ease in providing a berth for Banks' pet noble savage and he was consequently less than sympathetic about Tupaia's death, describing him as a shrewd, sensible, ingenious man, but proud and obstinate, 'which often made his situation on board both disagreeable to himself and those about him, and tended much to promote the diseases which put a period to his life'. The captain's attitude of making the punishment fit the crime would be revealed again before long with the death of valuable members of his own crew.

Banks continued to be very weak and the slaves employed about his rented house were not trained in tending the sick. Two Malay women were hired to nurse him and Solander back to health, 'Hoping that the tenderness of the sex would prevail even here, which indeed we found it to do for they turned out by no means bad nurses'. Very soon Solander became so ill that their physician, Dr Jaggi, applied mustard plasters to his feet, merely as a gesture because there appeared to be little hope of survival. Banks, who was receiving regular blood-letting for his own condition, faithfully sat with his companion throughout a traumatic night and, amazingly, by morning Daniel Solander had improved 'better than I had any reason to hope'. Dr Jaggi advised them to move out of the town to a house a couple of miles away on the banks of a briskly running stream and open to sea breezes, in country which reminded Banks of the low parts of his native Lincolnshire. Cook, remaining with his ship, heard of their plight and sent his own servant and a seaman to assist them, although he was also afflicted with malaria. The naturalists, together with Herman Spöring, the two *Endeavour* crewmen and ten slaves and local helpers thrived in the country, with Solander recovered enough to be able to walk about the house. Now the rainy season set in and their nights were disturbed by frogs in nearby ditches, 'ten times louder than the European ones', whose constant croaking became almost intolerable. Even here the unmerciful biting of mosquitos, which bred in every splash of water, was still to cause discomfort. Great sheets of torrential rain beat down on the house from the westerly monsoon and its roof and walls leaked so badly that, according to Banks, a stream ran through the dwelling with a force almost capable of turning a mill.

Out in the Batavia Roads the next to die was Charles Green's servant, John Reynolds, in the middle of December, followed by the able seamen Timothy Rearden and John Woodworth. The only member of the *Endeavour*'s entire complement who thus far appeared to be unaffected was the old sailmaker, John Ravenhill, an inveterate drunk, whose bloodstream was probably so infused with alcohol that no germs had a chance. The third Christmas away from home was looking bleak, until on 24 December Cook called everyone to the ship ready for the next stage of the journey. Banks went on board almost in a holiday mood, turning his back on Batavia and having nothing good to say about the local merchants who, he complained, 'have joined all the art of trade that a Dutchman is famous for to the deceit of an Indian. Cheating by false weights and measures, false samples, etc., are looked upon only as arts of trade'. He had grown to detest the authoritarian rule of the Dutch who were 'rather severe to the natives, and too lenient to their own countrymen'. Banks was never happier to take his leave of a place than this foetid swamp designed as an elegant colonial outpost, whose canals stank of human ordure and where dead horses, pigs and buffalos were left to putrify until carried away by a flood. James Cook saw Batavia as a place where Europeans need not covet to go, 'but if necessary obliges them they will do well to make their stay as short as possible otherwise they will soon feel the effects of the unwholesome air of Batavia which I firmly believe is the death of more Europeans than any other place upon the globe'. He was extremely worried that they had arrived with a healthy company and after a stay of less than three months, were leaving in the condition of a hospital ship. He discussed this with several Dutch captains in port and they were all of the opinion 'we had been very lucky and wondered that we had not lost half our people in that time'.

Cook took on 19 extra seamen to strengthen his depleted crew. The ship was now refurbished: freshly caulked, painted and provisioned, ready for sea. Everyone was anxious to get away on a Christmas Day that would follow none of the niceties of tradition observed on the previous occasions. There was a slight altercation when the Dutch authorities objected to one of the new sailors, an Irishman named John Marra, who was an itinerant, working when and where he could find a berth. The Dutch claimed he belonged to them and held Danish nationality, but Cook insisted he was a British subject and would not release the man. This brought accusations of Cook being 'ungrateful and discourteous' by the authorities, who previously had been most cooperative in assisting with repairs and provisions for the Endeavour. Cook stuck to his decision to keep Marra, although another of his crew was lost when the demoted Patrick Saunders deserted and disappeared into the maze of streets and alleys near the port. He had been under increasing suspicion for involvement in the affair of Orton's ears and, according to Parkinson, confirmed his participation by desertion to such a hellish hole when a reward was put up for anyone who could identify the culprit. After Saunders' disappearance, the reward remained unclaimed and the incident was closed. The search for him on Christmas Day 1770 meant they missed their sea breeze and had to wait until the following day for departure.

The colourful plants of Java were there for the picking in great profusion, but Banks and Solander had been too ill and it was left to the energetic Sydney Parkinson to collect what he could from Batavia and its hinterland. He helped to gather 151 species, from which 43 watercolours were eventually painted. Thirty of these were selected for engraving to form the final section of the Florilegium.

'There is not I believe a man in the ship but gave his utmost aid in getting up the anchor', wrote a weary Joseph Banks, 'so completely tired was everyone of the unwholesome air of this place'. A British Indiaman marked their departure with three cheers and 13 guns and soon afterwards the garrison fired a 14-gun salute, which was reciprocated by the Endeavour's gunners with their depleted armaments. They were on their way again.

8
Java — London
1771

Before their eyes in sudden view appear
The secrets of the hoary deep, a dark
Illimitable ocean, without bound,
Without dimension, where length, breadth, and height,
And time and place are lost...

John Milton (1608–1674)
Paradise Lost

THE new year arrived troubled by mosquitos, whose numbers had increased greatly since leaving Batavia. This seemed improbable at sea until Solander, standing on deck near one of the scuttle casks, which held drinking water, saw hordes of them breeding on its surface. The island of Krakatoa was passed slowly in the face of a strong wind and the thought of fresh sea breezes blowing away the last vestiges of the malaria generated a misguided sense of optimism in everyone, making them ill-prepared for what was about to happen. A week after leaving, Banks noted that almost all the ship's company was still sick or newly sick with either malaria or dysentery; he himself became extremely weak with excruciating bowel pains, Solander would not get up from his cot, and Spöring and Parkinson were so affected they were unable to do any botanical work. The water and supplies taken on at Prince's Island, their last anchorage, obviously carried the infection that led to what was known as 'the bloody flux' — dysentery, with its dreadful symptoms of severe inflammation of the bowels accompanied by uncontrolled diarrhoea and haemorrhaging. Over the following few weeks, the deaths would include Banks' secretary Herman Spöring and the draughtsman Sydney Parkinson. The formerly energetic young painter became part of a daily list and his burial at sea merited not even a passing mention in Banks' journal and only a brief note in Cook's. Banks was too weak, just managing to cling to his own life, to give way to any emotion about the loss of his faithful draughtsman. In the early morning of 29 January, during a week that saw 11 deaths, the astronomer Charles Green met his convulsive end.

At the end of January the company was so weakened that no more than a handful of sailors was capable of running the *Endeavour* and Cook was forced to reduce the watches to four men each. He wrote of the calamitous situation: 'having hardly well men enough to tend the sails and look after the sick, many of the latter are so ill that we have not the least hopes of their recovery'. It was not until the end of the first week of February that there was any indication of the disease easing. After the last deaths on 27 February, Cook said 'we are in hopes that they will be the last that will fall a sacrifice to this fatal disorder, for such as are now ill of it are in a fair way to recovering'.

With such an enfeebled crew, any violent or prolonged storms might have tested them beyond endurance leading to disaster; but only a sudden squall at the end of the month caused concern when the fore topsail was ripped across, forcing the *Endeavour* to bring to. Apart from that, the two-and-a-half month journey across the Indian Ocean was thankfully routine, with the occasional sighting of a Dutch ship en route to the East Indies and fair sailing in the trade winds. On 3 March they were nearing the southern coast of Africa and in the dark, foggy weather some of the crew thought they saw land in the far distance, but this was dismissed as impossible by the captain because all his observations indicated the coast must be at least 100 leagues away. However, at first light the next morning Cook discovered to his horror and amazement that they were heading straight towards land with all sails set because the undetected Aghulas current had carried the *Endeavour* far closer to Africa than dead reckoning showed. Following the dreadful realization that this might have meant the end of the voyage had their encounter been at night, Banks was rather scathing: 'Fortunate as we might think ourselves to be yet unshipwrecked we were still in extreme danger, the wind blew right upon the shore and with it a heavy sea which broke mountains high upon the rocks'. Some on board thought they could not avoid being driven ashore and it took four hours of standing off 'in the vicissitudes of hope and fear' before Cook had the ship moving to the south-west safely off the land and heading for the Cape. The most southerly part of the continent was skirted and they were forced to wait outside Table Bay for favourable winds before coming to anchor in the road at what was known to sailors the world over as the Cape Town. The old nautical saying, 'a passage perilous maketh a port pleasant' was particularly relevant at this moment.

The vessel's company was still in a sorry state, having 29 men incapacitated and barely more than six able to perform full shipboard duties, but they could now look forward to a period of quiet recovery in an infinitely more healthy climate than Batavia's. A British East Indiaman was due to be homeward bound and its captain sent across a gift of fresh fruit and vegetables because none of the *Endeavour*'s boats could be launched to get ashore in the face of high winds and very choppy seas. Conditions abated the next day and Cook, Banks and Solander were able to go into the town and take up lodgings, while the governor allowed a grant to rent a house for the ship's sick. Daniel Solander had been wretchedly ill ever since leaving Batavia and immediately was confined to bed, not interested in food and growing weaker each day. Banks noticed that his companion's normally rotund figure had become 'very much emaciated by his tedious illness'. He, on the other hand, was making a speedy recovery and his roving eyes got to work again and liked what they saw on taking stock of the ladies of the Cape. 'In general', Banks noted, 'they are handsome with clear skins and high complexions and when married (no reflection upon my country-women) are the best housekeepers imaginable and great child bearers; had I been inclined for a wife I think this is the place of all others where I could have best suited myself'. While Banks was enjoying the favours of female company again, Cook was busy getting letters written to the Admiralty and the Royal Society announcing safe arrival at the Cape, hoping to allay any fears that might have arisen about fate of the *Endeavour*.

After such a long time away there was now a feeling among the sailors of being within striking distance of home. The Cape Town welcomed a constant stream of ships from many nations drawn to what Cook called 'This great inn fitted up for the reception of all comers and goers'. Banks saw in this cosmpolitan community that: 'a far larger proportion of the population were real Dutch than those of Batavia. But as the whole town in a manner is supported by entertaining and supplying strangers, each man in some degree imitates the manners and customs of the nation with which he is chiefly concerned'. It was a well-run, civilized place with a temperate climate and plentiful local wine, although judged by Banks to be inferior to that of Europe. This allowed for rest, recovery and recreation for the *Endeavour*'s men and for Banks, indulgence in his favourite pursuits of observing and collecting, together with satisfying a renewed appetite for wine, food and sex after all those weeks of debility. He inspected the crops and the farm animals of the Cape with the eye of a successful farmer, collected plants, insects and the skins of wild animals, and various volunteers from the ship accompanied him on long and strenuous expeditions into the surrounding country.

Cook used most of his stay to make sure his crew was restored to health, arranging the provisions for the ship, supervising the necessary maintenance and, in general, doing everything possible to ensure the *Endeavour*'s successful completion of the final leg of her long journey. He also had plenty of time to think about the consequences of the voyage, now that it was nearing an end, and consider the results of his own actions, perhaps to justify the loss of human lives. He was fully aware that any other trip of more than a year's duration would have lost many more men to scurvy alone, but he imagined that the *Endeavour*'s toll was sure to be discussed in some detail by the London press 'and what is not unlikely, with many additional hardships we never experienced; for such are the disposition of men in general in the voyages that they are seldom content with the hardships and dangers which naturally occur, but they must add others which hardly ever had existence but in their imagination'. Cook went on in this vein line after line, with hardly a full stop to check his outpouring of personal doubts and fears. It was going to be interesting to see what reception the newspapers would give their arrival in England and whether the unexceptional death toll was really news.

Ten additional crewmen were signed on for the homeward run and after a month at the Cape the *Endeavour* weighed anchor on 14 April, setting out in light conditions for St Helena off the west coast of Africa. That afternoon the master, Robert Molyneux, died, deserving a better fate according to Cook: 'A young man of good parts but had unfortunately given himself up to extravagancy and intemperance which brought on disorders that put a period to his life'. The onward voyage was routine and uneventful; on 29 April the *Endeavour* passed the meridian of Greenwich and completed her circumnavigation of the globe before arriving at the island of St Helena, three days later. In the harbour was the 50-gun *Porland* under the command of Captain Elliot, together with the sloop *Swallow* guarding a convoy of twelve Indiamen soon to leave for England. Cook thought it would be a sensible idea to sail north with this fleet, although he realized that his lumbering ship with her new crew members, threadbare sails and chafed rigging would have a struggle to keep up. Captain Elliot generously provided firewood for the *Endeavour* as well as water and provisions while Banks and a still weak Solander went ashore to explore the area surrounding the British settlement, nestling under mountains rising sharply to 2,700 feet. According to Banks, the town standing beside the sea was generally ill-built with many of the public buildings close to ruin. The white inhabitants were almost all English and lived entirely by supplying ships with provisions, although there was not a great deal of variety in what they offered and a general lethargy pervaded their trading activities. 'Nature has blessed this island with very few productions either useful for the support or conducive to the luxury of mankind', was one of Banks' comments. It was just as well that their visit was brief because he found St Helena a wretched place and was highly critical of the British residents for their behaviour towards the islanders.

Cook knew that the British fleet would arrive home ahead of him and he gave Captain Elliot a letter for the Admiralty and more of the officers' logs and journals; he was anxious that London should have a full account of the *Endeavour*'s progress with additional information about the voyage since Batavia. He also alerted the Astronomer-Royal, Dr Maskelyne, to the sorry state of Charles Green's papers and the apparent anomalies in his calculations of the transit of Venus that arose because of the imprecise nature of the observations themselves. On the morning of 4 May they left with the fleet in an unusual and impressive spectacle — 14 ships leaving the harbour under full sail. In spite of Cook's fears about the *Endeavour*'s speed, she was able to keep up with the convoy and they were still together two weeks later when a surgeon had to be brought across from one of the Indiamen to examine Zachary Hicks, who had become dangerously ill. From this point the fleet gradually gathered speed and was almost lost in the distance when Hicks died on the twenty-fifth; it was the thirty-eighth death since leaving England. 'He died of a consumption', Cook wrote in his journal, 'which he was not free from when we sailed from England so that it may be truly said that he has been dying ever since, although he held out tolerably well until we got to Batavia'. Hicks' body was committed to the sea in the all-too-familiar ceremony and Charles Clerke was promoted to lieutenant in his place. By 27 May the *Endeavour* was on her own again with not even a distant sail on the horizon. Banks arose that morning and looked for signs of the British convoy but, 'to our great surprise we had no sight of the fleet, even from our mast head, so we were obliged to jog on by ourselves'. The weather continued fine and there was the occasional sail to give a hint that they were approaching the busy sea lanes to and from Europe. The *Endeavour* fell in with a schooner from Rhode Island engaged in whaling off the Azores and Cook sent a boat over for news which, after the gloomy information gleaned at Batavia and the Cape, was surprising: there was peace in England, just as when they had left, the dispute with the American colonies seemed to be over and there were no major conflicts on the continent of Europe. A few days later another schooner was encountered and this news was confirmed. Its master told of how he had chased a whale into the harbour of St Michael's Island and while pursuing it was fired upon by the Portuguese, forcing the hunters to leave their quarry to them. Cook arranged to purchase some fine salt cod and a supply of New England rum but he was unable to grant a request for beef because there was almost none left in his stores. The British convoy came into sight again after the ship had made up some speed, although her sails were in such a pitiful state that Cook reported, 'some thing or other is giving way every day'. Because of the need to put as little strain as possible on the canvas and rigging, the *Endeavour*'s spread of sail was now reduced, cutting back her speed, so that the British ships pulled away and vanished over the horizon.

It was the best time of the year for a square rigger to be making this journey, during the summer with reliable breezes to take her all the way to the English Channel. In fine sailing weather and with a healthy crew, the ship was given a final clean and polish as the homesick sailors went about their tasks with renewed enthusiasm. Masts and spars were scraped and varnished, the deck scrubbed and cleared of most evidence of a long occupation by live animals. Blocks and tackle were overhauled and the never-ending work of patching the sails continued. By early July other vessels were constantly in sight and it was learned from a passing British brig bound for the West Indies that wagers had been laid in London on the chances of their survival. In fact, there had been much newspaper speculation over the last eighteen months and, in particular, *Bingley's Journal* of the previous September concocted an interesting theory about their fate against the background of rumours that Spain and Britain were close to war. The story read: 'It is surmised, that one ground of the present preparations for war, is some secret intelligence received by the Ministry, that the Endeavour man of war, which was sent into the South Seas with the astronomers to make the observation, and afterwards to go into a new track to make discoveries, has been sunk with all her people, by order of a jealous Court, who has committed other hostilities against us in the Southern Hemisphere. Mr Banks and the famous Dr Solander, were on board the above vessel, and are feared to have shared the common fate with the rest of the ship's company'. Much of this was speculative nonsense, although the story did reveal that Cook's confidential instructions from the Admiralty were not so secret after all, at least to London journalism. The item had caused great concern among Banks' family and friends, including Harriet Blosset, although Banks' strong-willed sister, Sarah Sophia, entertained few doubts about her beloved brother's safety. In early October 1770, soon after the story appeared in *Bingley's Journal*, she wrote to the natural historian Thomas Pennant in Flintshire stating there was 'not the least foundation for the very alarming reports that were in the newspapers regarding the *Endeavour*'. She thought that her brother might not be home until the spring of 1771 because of missing the trade winds, 'but that is a very different situation from what the papers reported'. Fears for their safety were allayed by January 1771 when the *General Evening Post* reported the safe arrival of the *Endeavour* at Batavia. In May the *London Evening Post* printed an extract from a letter, also sent from Batavia, by Sydney Parkinson. London awaited the return of the adventurers with growing excitement and Parkinson's family was delighted to learn from the *Public Advertiser* of 21 May: 'We are told that Dr Solander, Mr Banks, and his draughtsman, Mr Parkinson, were all well at St Helena the latter end of February'. Sadly, the artist had been dead for four months.

The 'famous' Dr Solander was making a slow recovery from his illness and regaining some of his former ample bulk. Banks was ebullient about getting home and being able to show off his vast collection of curiosities from the South Seas, but he experienced a sad moment during these final days at sea when Lady, his greyhound bitch, died quietly on her favourite seat in the great cabin. The scourge of Tahitian hogs, New Zealand rats, New Holland kangaroos and Batavian monkeys was committed to the ocean. On 10 July soundings indicated they were approaching the Scilly Isles and then Nicholas Young at the masthead sighted the coast 'which we judged to be about the Land's end', according to Cook. Everyone rushed on deck in the fresh, clear weather to watch England loom ever closer. Cook cast only a cursory glance at this familiar coastal profile as he was busy writing a final shipboard letter to Mr Stephens at the Admiralty. 'You will herewith receive', he penned, 'my journals containing an account of the proceedings of the whole voyage together with all the charts, plans and drawings I have made of the respective places we have touched at, which you will be pleased to lay before their Lordships. I flatter myself that the latter will be found sufficient to convey a tolerable knowledge of the places they are intended to illustrate and that the discoveries we have made, though not great, will apologise for the length of the voyage'. Excruciatingly modest and self-effacing to the end, Cook made preparations to leave the ship together with Banks and Solander. At noon on 12 July the *Endeavour* passed Dover and shortly after anchored in the Downs. The intrepid trio disembarked at Deal under the summer sun, and, after taking lodgings for the night, travelled by coach to London the next day. They had been absent from England for 1051 days.

In a postscript to his journal Cook was already anticipating further discoveries in the Pacific, suggesting that the best way of entering the vast region was from New Zealand, 'First landing and refreshing at the Cape of Good Hope, from thence proceed to the southward of New Holland for Queen Charlotte's Sound'. By leaving this watering place by late September or early October at the latest, the whole southern summer could be made available for discoveries before running eastward 'in as high a latitude as you please, if you met no lands', which would leave time to round Cape Horn before the winter storms and return home through the Atlantic. If land was encountered and the passage delayed, it would be possible to go back westward by way of Tahiti to make the discoveries in the South Seas complete. Cook, Banks and Solander talked about this on the journey to London and agreed that further exploration must be made quickly in order to stay ahead of the French. Banks was prepared to commit himself to another adventure almost immediately, but Solander knew there was a huge amount of work to be done if the botanical results of their three-year voyage were to be put in order and published for the benefit of science. Cook was non-committal about his participation in further exploration. He would soon be 43 and had to follow a career according to the Admiralty's regard of his conduct on the voyage just completed. He had shouldered the main burden of difficult decision-making over the long period of his first command and he needed time away from the sea to return to his home, wife and family before facing whatever his superiors next expected of him. Daniel Solander, the 37-year-old bachelor, had no such ties except a life-long involvement with botany. Joseph Banks now aged 28, relishing his freedom, was in no mood to consider conventional domestic constraints. It seemed inevitable, however, that whatever happened to each of them in the short term, their destinies would somehow be linked in the future. Banks had begun the *Endeavour* voyage regarding the captain as little more than a talented ship driver, while Cook saw the gentlemen as unwelcome extra bodies, effete landlubbers who wished to know nothing about the sea or sailing. Three years of close confinement together and many harrowing moments had changed all that. There had been a remarkable lack of serious personal dispute. Although they were from entirely different backgrounds at a time when privilege was paramount, Cook and Banks now held a healthy respect, even affection, for each other, which augured well for a continuing association.

PART TWO

BANKS' FLORILEGIUM

9
Beginnings
1771–1820

We have ploughed the vast ocean in a
fragile bark.

Ovid (43BC–AD18)

THE capital learned of the explorers' return from an item in the *London Evening Post* of 15 July 1771: 'On Saturday last an express arrived at the Admiralty, with the agreeable news of the arrival in the Downs of the *Endeavour*, Captain Cooke, from the East Indies'. It was barely two months since the same paper had reported the vessel's safe arrival at Batavia. The story continued: 'The ship sailed in August 1768, with Mr Banks, Dr Solander, Mr Green, and other ingenious gentlemen on board, for the South Seas, to observe the transit of Venus; and have since made a voyage around the world, and touched at every coast and island where it was possible to get on shore, to collect every species of plants and other rare productions in nature'. It was the last time the leading personalities of the voyage would receive equal billing; at first, the popular press made Solander their star explorer, probably because of the 'Dr', and then Mr Banks. The captain, whose name was often misspelt, received diminishing attention. The report added: 'Their voyage, upon the whole, has been as agreeable and successful as they could have expected, except the death of Mr Green, who died upon his passage from Batavia. Dr Solander has been a good deal indisposed, but it is hoped a few day's refreshment will soon establish his health. Captain Cooke and Mr Banks, we have the pleasure to inform the public, are perfectly well'. John Ellis informed Linnaeus in Sweden that the botanists were back, 'laden with the greatest treasure of natural history that ever was brought into any country at one time by two persons'. In reply, the master implored Ellis to send 'some specimens of plants from Banksia in Terra Australis for the new-found country ought to be named Banksia from its discoverer, as America from Americus'.

James Cook returned to his family's modest little dwelling on the Mile End Road in east London, Daniel Solander repaired to his bachelor apartment and Joseph Banks hurried back to the bosom of his own family. His devout and devoted mother and sister welcomed him, praising God for affording protection through so many vicissitudes. Sarah Sophia confided to

125

her diary: 'How often have my dear mother and self contemplated and admired the innumerable and grave dangers he has escaped; and adored our gracious God, for restoring him to us'. Their conquering hero was reluctant, however, to honour the pledge made three years before and make contact with Harriet Blosset, who remained patiently at her country retreat. In fact, nothing was further from Banks' mind. He now regarded that affair as a carefully calculated plan to snare him and he felt that at this stage of his life marriage and science, in the form in which he wished to practise it, were incompatible. Harriet waited in vain while her former lover basked in a glow of growing adulation. A week after the *Endeavour's* return, the *London Evening Post* and the *General Evening Post* vied for stories about the exploits of Banks and Solander. They reported the discovery of 'a Southern Continent in the latitude of the Dutch Spice Islands' to which, as a consequence, 'more ships will be destined in search of this new terrestrial acquisition'. There was also the announcement that Dr Solander and his party 'brought over with them above a thousand different species of plants, none of which were ever known in Europe before'. By the end of the month the papers were clamouring for additional colourful background news of the voyage, but only sensational gossip was available because Cook had collected his officers' journals at Batavia and the Cape. These documents, together with his own, Solander's and Banks' writings, were handed over to the celebrated littérateur, Dr John Hawkesworth, who was to prepare a comprehensive account of recent British South Seas exploration with the intention, in his own words, of 'bringing the adventurer and the reader closer together'. He was a Londoner who had risen from humble beginnings and a slender education to become a member of the capital's literary elite. With the help of his wife's fortune he had also been a director of the Honorable East India Company. He was recommended to Lord Sandwich for 'writing up' the official account of the *Endeavour* voyage by Dr Charles Burney, the musician, and the actor David Garrick. Dr Hawkesworth accepted the very handsome fee of £6,000 for his services, which was considerably more than the purchase price of the *Endeavour* itself.

The plant material collected and sorted on the voyage was extensive, with the herbarium specimens accounting for about 110 new genera and 1,300 new species, while Parkinson's efforts provided the basis for nearly 1,000 illustrations. Banks and Solander had carefully classified their plants according to the methods of Linnaeus, who wrote to his friend John Ellis: 'I have every day been figuring to myself the occupations of my pupil Solander, now putting his collection in order, having first arranged and numbered his plants, in parcels, according to the places where they were gathered, and then written upon each specimen its native country and appropriate number. I then fancied him throwing the whole into classes; putting aside, and naming such as were already known; ranging others under known genera, with specific differences; and distinguishing by new names and definitions such as formed new genera with their species. Thus, thought I, the world would be delighted and benefitted by all these discoveries; and the foundations of true science will be strengthened, so as to endure through all generations'. But now the thought of starting the exacting task of organizing the botanical material was pushed aside by the heady atmosphere of triumphant return.

Bingley's Journal of 27 July revealed that the ship 'lost by the unhealthiness of the climate, 70 of her hands, though they were picked men and had been several times in the Indies. However, those who survive will have made their fortunes by traffic, having brought home some of the richest goods made in the East, which they are suffered to dispose of without the inspection of Customs House officers'. The newspapers were persistent in their quest for *Endeavour* anecdotes and they managed to acquire a couple of what were described as personal memoirs, together with a bizarre account of Charles Green's demise. This stated that he had been ill for some time 'and was directed by the surgeon to keep himself warm, but in a fit of phrensy he got up in the night and put his legs out of the portholes, which was the occasion of his death'. The report added that Green's observations of the transit were complete and intact. The *General Evening Post* gave notice on 27 July, just a fortnight after the *Endeavour*'s return, of an authentic account of the natives of Tahiti 'together with some of the particulars of three years voyage made by Mr Banks and Dr Solander'. The journal claimed to be printing a copy of an original letter sent by a person on the ship, but no identity was revealed. It proved to be a ponderous account of the journey, spiced with some titillating information about Tahitian women and girls, who were said to marry at nine or ten, bear many children and at the age of 22 become old and ugly. Young virgins were readily available for the sailors' dalliance at 'three nails and a knife' and the writer boasted that he was 'the buyer of such commodities'. He also revealed they had an *outré* fashion, these 'nut-brown sultanas', of painting their bottoms jet black. Amours took place with little regard for time or place and their dances were most 'indecent', incorporating many 'obscene gesticulations...like the Indostan dancing girls'. The *Public Advertiser* told its readership in early August that the *Endeavour* 'Sailed many hundred leagues with a large piece of rock sticking in her bottom, courting destruction had it fallen out'. Not to be outdone by such sensationalism, the rival *London Evening Post* printed what purported to be the copy of a letter about Tahiti from a gentleman on board the *Endeavour* which stated: 'Monsieur Bougainville had been here before us with two sail of ships, and brought the French disease among the poor people'.

With press notices like these, the returned voyagers could hardly fail to be the talk of the town and they quickly became celebrities, very much in demand at society gatherings. The President of the Royal Society, Sir John Pringle, introduced Banks and Solander to the king at Windsor and this formal meeting was soon followed by a more relaxed assembly at Richmond Lodge, where the botanists were able to describe their adventures and discoveries in detail to the sovereign, who was fascinated by what he learned. Banks and Solander were joined by James Cook on a visit to Lord Sandwich at his country estate, Hinchingbrooke. Banks had been elected in his globetrotting absence to the Royal Society Club, an exclusive dining fraternity of philosophers and scientists. In November he went to Oxford to receive with Solander the honorary degree of Doctor of Laws. The adventurers were seen in all the best places: 'The most talked of at present', Lady Mary Coke noted, 'are Messrs. Banks and Solander. I saw them at Court and afterwards at Lady Hertford's, but did not hear them give any account of their voyage around the world which I am told is very amusing'. Conversations about their travels did tend to the fatuous and eventually became downright dithyrambic. An evening at the house of Sir Joshua Reynolds saw them in the company of Dr Samuel Johnson and his biographer James Boswell, who asked the good doctor if he had not had some desire to go upon the *Endeavour* expedition. 'Why, yes', was his reply, 'but I soon laid it aside. Sir, there is very little of intellectual in the course. Besides, I see but at a small distance. So it was not worth my while to go and see birds fly, which I should not have seen fly; and fishes swim, which I should not have seen swim'. Boswell pointed out to the sage that the principal reason for the gentlemen's long voyage was, in fact, the study of new plants. This evening in the company of Dr Johnson resulted in his tribute, not to the great navigator who had made the circumnavigation possible, nor for once to Banks and Solander, but to the much-travelled nanny-goat, which had kept the gentlemen in fresh milk for the entire journey. It read, when translated from its original Latin:

In fame scarce second to the Nurse of Jove
This goat, who twice the world has traversed round,
Deserving both her master's care and love,
Ease and perpetual pasture now has found.

The gentlemen thought the industrious animal probably deserved a lush pasture much more than a limp Latin verse, although Dr Johnson did have the good grace to suggest to Banks in the accompanying letter: 'You, Sir, may perhaps have an epic poem from some happier pen than mine'.

Banks might have wished for a little ease of his own and the pleasures of his Lincholnshire meadows, but his treatment of Harriet Blosset led to a good deal of emotional turmoil. Having waited a decent time, and hearing nothing from him, Miss Blosset finally made the journey to London to broach the subject of their marriage. There were emotional scenes when she learned that Banks had lost all interest in their arrangement. Then like a passing summer storm, he had a sudden change of heart, only to return to his previous stand of refusing to entertain matrimony. If Joseph Banks had wanted to reveal the truth as a man of honour he would have admitted to her that he was being advised against the match by eternal bachelor Daniel Solander and, probably more to the point, he had fallen in love again since returning to London. Known only to the wider world as Miss B — n, his new sweetheart was described by a close observer as 'genteel', with particularly engaging looks: 'All the elegant accomplishments were united in her and were only surpassed by her mental improvements'. Banks had known her since she was a beautiful 17-year-old. During the *Endeavour* voyage her father had died, having lost a fortune in gambling debts and leaving his daughter penniless. She was rediscovered living as companion to an elderly lady. Banks started visiting her and she soon became his mistress, installed in a house at Orchard Street off Portman Square. She bore a child by him, but what happened subsequently to mother and baby remained a closely guarded secret. Banks extricated himself from the promises he had made to Harriet Blosset by the payment of a handsome settlement. As one commentator put it, he finally made amends for all the waistcoats she embroidered while he was sailing around the world.

Having bought his way out of Miss Blosset's life, Banks thought he was free to enjoy the pleasures of his new liaison. There would also now be time to organize the plant collections and initiate the publication of the botanical material from the *Endeavour* voyage, which science was eager to know about. The naturalist John Ellis had written to Linnaeus complaining 'they are so busy getting their things on shore and seeing their friends that they have scarce time to tell us of anything but the many narrow escapes they have had from imminent danger'. To which the great man replied, 'I cannot sufficiently admire Mr Banks who has exposed himself to so many dangers and has bestowed more money in the service of natural science than any other man. Surely none but an Englishman would have the spirit to do what he has done'. A grand folio work in many volumes was planned as a comprehensive illustrated record of the flora from the principal areas visited and a considerable sum of money was earmarked by Banks for the project. Solander was now officially Banks' secretary/librarian and custodian of his collections, although he also worked for the British Museum and was engaged to arrange the Duchess of Portland's extensive shell collection. However, with affairs of the heart apparently settled, another distraction from his scientific pursuits suddenly confronted Banks.

Sydney Parkinson's will had been lodged with his sister Britannia before he left England. Soon after the *Endeavour*'s return, Parkinson's elder brother Stanfield, an upholsterer with an erratic temperament, became suspicious that some of his brother's papers and artifacts were being withheld. The document bequeathed the young draughtsman's salary, of which £160 was owing, to Britannia, while Stanfield was given possession of all the other effects. Under normal circumstances the captain of the ship would have been responsible for these matters but the situation was complicated by Banks being the artist's employer and, therefore, the artist being his personal responsibility. Stanfield called at New Burlington Street and Banks told him with the privileged air of his class that he was extremely busy in settling personal affairs after the long voyage and an account of the dead man's effects would be forthcoming as soon as he could find time to attend to it. Stanfield understood that a journal was bequeathed to James Lee, the Quaker nurseryman of Hammersmith, and he wanted to know if this was a fact. He was keen to acquire the journal himself because of the rapacious demand from newspaper editors and book publishers for any sort of material about the *Endeavour* voyage. Parkinson's extensive writings were worth a good deal of money in this market. Several weeks passed and nothing happened about the effects. Stanfield returned to New Burlington Street, but the botanist was in no mood for confrontation, having experienced too many of late. He sent Stanfield away again with a vague promise to settle the business in his own good time. It was merely one of several problems at the moment, although for Stanfield the affair had become an obsession.

Stanfield Parkinson was persistent. He began seeking information from the *Endeavour*'s officers about what hapened at the time of his brother's death. The details he learned confirmed his suspicions that Banks was procrastinating, because a number of drawings said to have been done in his brother's own time were missing. Stanfield was informed that immediately after Parkinson died, Banks and Solander went to the cabin with the captain's clerk, Richard Orton, and made an inventory of the effects, although Orton was now of the opinion that it was inaccurate because of the failure to include all the young man's possessions. Another five weeks passed and then Stanfield received a message from Banks asking him to come and settle his brother's affairs the next afternoon. Banks had been informed of the inquiries among the crew and felt compelled to guard his own interests by engaging a lawyer. With his legal representative in attendance, he informed Stanfield that he was claiming the right to all of Parkinson's journal notes and drawings as his employer. Although Solander had admitted that the dying artist had asked him to allow James Lee to inspect his papers, Banks said this was just a passing mention, there was no written confirmation of it. Banks then produced a bundle of papers which Stanfield recognized as being in his brother's hand: they were the drafts of the journal Sydney Parkinson was working on throughout the voyage and which several of the ship's officers had thought was the best description of the *Endeavour*'s journey.

Stanfield countered this latest move by seeking the assistance of a fellow Quaker and noted botanist, Dr John Fothergill, who was also a member of the Royal Society and a leading Westminster physician. He advised moderation, preferring to rely on the inherent gentlemanly attitudes of Joseph Banks to settle the difference. Stanfield, however, wanted quick satisfaction and urged Fothergill to make a direct approach to his fellow Royal Society member and also to Daniel Solander, suggesting a suitable payment might

Banks' mistress known only as Miss B—n. A contemporary engraving.

Kangaroo by George Stubbs, 1771–72. Painted from a skin taken back to Britain on the *Endeavour*.

bring the situation to a rapid conclusion. Banks, anxious to be done with it, wrote to Fothergill accepting his suggestion that £500 including the £160 of back salary, would settle the account. This was accepted by Stanfield and Britannia Parkinson, who met Banks at the end of January 1772 to sign a receipt for the money, with Dr Fothergill as witness. There were, however, some loose ends in the agreement, including the value placed on the dead draughtsman's collection of sea shells, which was in Stanfield's possession and was being re-negotiated by Banks. Stanfield was also very persistent in wishing to see his brother's papers, which Banks finally allowed him to do. By now the older brother was convinced that he had been outsmarted and immediately had the journal drafts copied and prepared for publication by a Grubb Street hack journalist, James Kenrick, who was notorious for his libels. When news of this reached Dr Hawkesworth, whose official record of the voyage was still more than a year away from completion, he applied to Chancery for an injunction against the publishing of Parkinson's journal, which was initially successful.

Attempts to confine accounts of the *Endeavour* voyage to Hawkesworth's forthcoming publication were partly foiled by the barrage of newspaper reports and an anonymous journal that had appeared in September 1771, only weeks after the vessel returned. It was a slim volume, the title of which seemed almost to rival the bulk of its contents, carefully detailing what the reader could expect inside:

> *Journal of a voyage round the world in His Majesty's ship Endeavour, in the years, 1768, 1769, 1770 and 1771 undertaken in pursuit of natural knowledge at the desire of the Royal Society, containing all the various occurrences of the voyage, with descriptions of several new discovered countries in the Southern Hemisphere, and accounts of their soil and productions and of many singularities in the structure, apparel, customs, manners, policy, manufactures and of their inhabitants.*

An added curiosity was the inclusion of *A concise vocabulary of Otaheite*. There was immediate speculation about the author's identity and the names of Banks, Solander and Cook were canvassed, although it was generally known that their journals and writings had been handed over to John Hawkesworth. The most likely candidate for what was obviously no fabrication was the young man who had been so disliked by Cook, James Mario Matra. He had signed on at the start of the voyage at Deptford as an able seaman but was accepted as a kind of cadet with access to the quarter deck and became a midshipman during the final stages of the voyage. The master roll had shown him as Magoa, although he was also known as Magra. This name came from his family, of Corsican background, who lived in New York, where James was born about 1748, and where his father had become a fashionable physician. The published journal contained only highlights of the voyage, concentrating on the time spent in Tahiti, but it satisfied the demand for an extended account of the journey of discovery, selling out its first edition with an immediate reprinting and a translation into French that appeared in Paris within a year.

The *Endeavour* voyage was a huge popular success, although for reasons other than its instigators might have intended. Neither the observations of the transit of Venus nor the search for the great southern continent had produced anything of obvious value and Cook found it necessary to be apologetic to the Admiralty; in fact, virtually grovelling to them for having discovered so little that was new. The Navy Board, however, was more than satisfied with his superb surveys of Tahiti, New Zealand and New Holland, and a second expedition was proposed to continue the work of exploration where the last one had ended, possibly even to the South Pole. Lord Sandwich suggested it should have the successful basic team of Cook, Banks and Solander but this time on a more lavish scale. Two vessels would go to the far south, both cat-built barks, and Banks' entourage would be expanded officially to 16 men, together with one unofficial woman. She would secretly join him at Madeira, in the spirit of Bougainville's botanist, whose valet turned out to be a female. Philibert Commerson's lady, Jeanne Baret, dressed in male clothing and wearing her hair in a boyish style, had deceived most of the ship's company, although the fact that the valet was sleeping with his employer did seem a trifle strange. It was not until *La Boudeuse* was at Tahiti that the servant's sex became general knowledge on board when the natives were quick to see through the disguise and recognize a *wahine*. Banks was arranging for an attractive young lady from Edinburgh, named Miss Burnett, to disguise herself as a man, go to Madeira, and await his arrival.

The second voyage would be on the *Resolution*, formerly the Whitby-built *Marquis of Granby* of 462 tons, about a quarter as large again as the *Endeavour*. Her configuration, however, was similar, which meant that extensive additions and alterations needed to be carried out to accommodate the passengers and special plant cabins had to be constructed on deck. The small cabins with their low bulkheads were to be expanded and the between-decks spaces entirely revised. Banks assumed the authority to supervise these changes because of the fortune he was investing in the project and the backing he could expect from his close friend the Earl of Sandwich, who also happened to be the First Lord of the Admiralty. James Cook, who had not received the promotion to captain he could have expected after the *Endeavour* voyage, now held the rank of commander, which meant that he might never become a full captain. However, he was prepared to assist Banks wherever possible to make additions for comfort, including a stove for his cabin, green baize floorcloth in the great cabin and special brass hinges and door locks to add a touch of elegance. The *Resolution* was to have a companion vessel of 340 tons, the *Adventure*, to be commanded by Lieutenant Tobias Furneaux. At first the naval surveyors working on the ships said little could be done about additions because they were designed as rather basic cargo carriers and if their structures were raised they would become wildly unstable. Banks arrogantly rejected this and wrote to Lord Sandwich demanding the appropriate additions, which were carried out, but a trial run of the *Resolution* soon proved the surveyors right. Charles Clerke, the young man who had been promoted to officer on the *Endeavour*, thought the changes made the *Resolution* the most unsafe vessel he had ever heard of. He said he was prepared to go to sea in a grogtub if required, but this top-heavy craft was unlikely to survive a journey down the Thames, let alone a voyage to the South Pole. Banks would not be thwarted, the sentiments he had expressed in a letter to a friend at the end of 1771 still held sway: 'O how glorious it would be to set my heel upon the Pole! and turn myself round 360 degrees in a second'. Cook managed to keep out of this growing dispute. Wisely, he remained at home completing his charts and surveys from the *Endeavour* voyage and preparing them for publication.

Banks appealed once again to Lord Sandwich for assistance, threatening to withdraw himself and his party from the proposed voyage unless satisfactory alterations were carried out. He heard rumours that Solander was being asked to go with or without his benefaction and, noting what a darling of society he had become with his genial personality, Banks became even more resolute in getting his own way, complaining that over-crowding in the great cabin would make impossible the labours of the men of science and the accompanying artists. As this correspondence about the inadequacies of the accommodation on board lengthened, the strain on the friendship with Lord Sandwich, who was admirably controlled in his reaction to Banks' demands, increased. However, the Navy Board chose to be more direct about his despotic manner: 'Mr Banks seems throughout to consider the ships are fitting out wholly for his use; the whole undertaking to depend on him and his people; and himself as director and conductor'. Banks' next demand was for two fast frigates instead of the Whitby ships and the board countered the suggestion by pointing out that if this type of vessel had been in the *Endeavour*'s place on the New Holland coast it would never have been heard of again. Finally, Banks had to accept that the sheer weight of naval bureaucracy was getting the better of him and he ought to withdraw from the project with dignity before his reputation became tattered. While all the wrangling was going on he had spent more than £5,000 on stores, appliances and presents for natives, and he continued to employ a large staff on considerable salaries. He felt the need to get away and thought it prudent to employ his staff 'in some way or other to the advancement of science'. Remembering the productive time he spent some five years before in Newfoundland and Labrador, Banks, prompted by Solander, now became interested in the prospect of a trip to Iceland. It might produce some interesting botanical and zoological specimens and the study of active volcanos could also be rewarding.

A year to the day after the *Endeavour*'s return, Banks and Solander were off again, sailing down the Thames on a 190-ton brig chartered for £100 a month. At the same moment the *Resolution* was slipping out of Plymouth Sound, bound for the South Seas. Her first stop was to be Madeira, where Miss Burnett was waiting for her lover. Word finally reached her that he was not on board a few days before the ship's arrival. Cook learned that a person named Burnett had just left the island after waiting for nearly three months for Mr Banks. Cook was told that every part of Burnett's behaviour and every action had suggested the ways of a woman: 'I have not met with a person that entertains a doubt of a contrary nature'. The *Resolution* and *Adventure* were expected to be away for at least two years, but Banks planned his journey to last only four months, a summer excursion.

Iceland was reached at the end of August and Solander soon discovered he could converse in Swedish with the locals, who Banks found to be of a good honest disposition, but also leadenly serious and sullen. They were fascinated with the geysers in an area where volcanic activity had otherwise been dormant since the violent eruptions of a few years before. Banks, always keen for gastronomic experiment, supervised the cooking of salmon trout and a piece of mutton in a hot spring and on another occasion noted that ptarmigan took only six minutes to be ready for eating by this method. The collecting became paramount; everything from plants, birds, fish and insects to ancient Icelandic manuscripts, which were destined for the British Museum. A 12 day expedition was mounted to scale the active volcano, Mount Hecla, and they were the first to climb through the ice and snow to reach its summit. After a month in Iceland the season was getting late and it was time to return home with a load of volcanic rock as ballast, some of which was destined for Kew and the remainder to form a decorative rockery in the Chelsea Physic Garden. Banks extended his trip by staying in the Scottish Highlands on his way back before repairing to London, where he would sort the Icelandic collections with Solander and begin preparing for publication of the *Endeavour* folios. He had virtually expunged the travel bug from his system and, although he never ceased to dream of colourful adventures on distant shores, his grand touring was effectively over at the age of 29.

After Cook had left for the Pacific, John Hawkesworth's official record of the *Endeavour* voyage was published at the price of three guineas with the title:

An Account of the Voyages undertaken by order of his present Majesty for making Discoveries in the Southern Hemisphere, drawn up from the

Journals which were kept by the several commanders and from the Papers of Joseph Banks Esq.

The three-volume work, with its first section devoted to the voyages of Byron, Wallis and Carteret in the *Dolphin* and *Swallow*, proved to be a verbose disappointment, although a lively seller, quickly running to a second edition, together with versions in Dublin, New York, Paris, Rotterdam and Berlin. The epic journey of Cook and Banks, based upon the records of those two men, was made to appear rather dull because its author, an armchair traveller, transposed events into his own style and presented them in the first person singular, embellished with 'such sentiment and observations as my subject should suggest'. The dull thud of Hawkesworthian prose, based not a little on the style of his friend and former collaborator, Dr Johnson, buried the individual personalities until the reader was unsure whose voice was being heard at any time. Certainly Daniel Solander and

Sydney Parkinson from the frontispiece of his journal published in 1773.

Joseph Banks did not receive just attention in the narrative and the author showed his feelings about Stanfield Parkinson's rival publication by failing to give a single mention to Sydney Parkinson or acknowledge his many drawings, which were the basis of the book's illustrations. Inaccuracies, fantasies and evasions abounded to the point where the natives of Tierra del Fuego, originally described by James Cook as 'Perhaps as miserable a set of people as are this day upon earth' became, when filtered by Hawkesworth, idealized noble savages free 'from care, labour and solicitude'. The work was a huge popular success, however, in spite of the rather dismissive reviews from critics such as Horace Walpole: 'The entertaining matter would not fill half a volume; and at best is but an account of the fishermen on the coast of forty islands', and Dr Johnson: 'Hawkesworth can tell only what the voyagers have told him; and they have found very little, only one new animal, I think'. The Parkinson journal, which had been restrained by Hawkesworth's injunction, was also published in 1773 as *A Journal of a Voyage to the South Seas in H.M.S. Endeavour, faithfully described from the papers of the late Sydney Parkinson*. In a long preface written for him, since Stanfield was illiterate, he related his side of the dispute with Banks. Both men responsible for these *Endeavour* publications would not live long enough to enjoy kudos or royalties from their labours. Dr John Hawkesworth died in November the same year, persecuted by 'those envious and malignant witlings of the literary world', according to Fanny Burney, and was buried near his home at Bromley in Kent, while Stanfield Parkinson was condemned to an insane asylum, where he died a few months later.

Banks was never particularly interested in politics or the law; if he had been, the Stanfield Parkinson affair might have been settled in a different way. His main interests, apart from drinking, wenching and good conversation — usually in moderation — were agriculture and science in general, and botany in particular. He was most at ease in scientific, literary and artistic circles, frequently seeing such notables as Samuel Johnson, Joshua Reynolds, who painted his portrait, the zoologist Thomas Pennant, Garrick the actor and Lord Sandwich. Other friends were often connected to the Banks family by marriage, and plays, operas, concerts and masquerades, together with fishing parties in the company of ladies of pleasure, were also a prominent feature of his busy life. The wealth that flowed regularly from the Revesby estates allowed him to be a true dilettante, a gentleman amateur, doing as he wished and going where he wanted with no financial constraints. These attributes were attractive to King George, who was only five years older than Banks, and they developed a close friendship based on the spirit of their mutual enthusiasms. One of these was the great royal gardens at Kew, which Banks described as 'the finest botanical gardens in Europe' and, although he travelled very little on the continent to see for himself, he was not far wrong. With the king's patronage Banks succeeded Lord Bute in the honorary position of director of the gardens during 1772 and in the years ahead would be responsible for the introduction of many new plants into England. It was part of a plan to make Kew the great hub of the growing Empire's botanical pursuits, along with a continuous interest in improving the gardens. Classification and study would be part of the scheme, but the acclimatization of plants from one part of the world for growth in another was also central to their thinking, and experiment needed to be viewed in the light of what was commercially practical. Eventually, collectors would be appointed to go out on naval expeditions to gather specimens for Kew, and later the traffic would be reversed as cases of seeds and plants were sent out from England to infant colonies.

Banks himself travelled only once more outside Britain, in March 1773, when he went with the Hon. Charles Greville to attend a meeting of the Batavian Society at Rotterdam. They also visited Leyden, Haarlem and Utrecht in that 'fenny, muddy country', as Banks, a fenman himself, described Holland, and paid their respects to the Prince of Orange at the Hague. Banks had prepared a long discourse 'On the Manners of the Women of Otaheite' to amuse the Princess. It was a lighthearted and rapturous account brimming with enthusiasm for the South Seas society where love was the chief occupation and only in complexion were European ladies superior to the wonderful goddess-like creatures from the Pacific. Superlatives ran on line after line for several pages about the genuine innocence and modesty displayed by the Tahitian maidens baring their breasts in a country where, according to Banks, chastity was esteemed as

Caricature of Solander published in London, 1772.

virtue and women were no less inviolable in their attachments than in Europe. He admitted there was a nasty practice of infanticide among the natives but it was, he explained, due largely to the barbarity of the men. The ladies, who might have come from the chisel of a Phidias or the pencil of an Appeles, excelled beyond all compare with other nations in cleanliness in this ideal land of liberty. Such comments left Banks open to the barbs of London satirists and already he and Solander had been on the receiving end of a couple of rather sharp caricatures. These had related to Macaronis, a name originating with Italian pasta then fashionable in England and defining a person who exceeded the normal bounds of lifestyle, usually connected with luxury or extravagance; in other words a fop. *The Fly-Catching Macaroni* was a hand-coloured etching representing Banks with the ears of an ass, standing astride two hemispheres attempting to catch a brightly-coloured butterfly, with the text:

Caricature of Banks published in London, 1772.

I rove from pole to pole. You ask me why.
I tell you Truth, and catch a —— Fly.

The Simpling Macaroni was a picture of Solander holding a plant in one
hand and a naturalist's knife in the other:

Like Soland Goose from frozen zone I wander,
*On Shallow Banks grow fat, Sol*****.*

Hawkesworth's publication led to a flood of verse satires in London, many
of which concentrated on the apparent promiscuity of Tahitian women, the
insatiable sexual appetite of Queen Oberea, together with the moral lapses
of the *Endeavour*'s crew in general and Joseph Banks in particular. He
became the central figure in an outpouring of outrageous lampoons and

Banks introduces Omai to Solander.
Painting by William Parry, c1776.

salacious satires, like this one, part of *An Epistle from Mr Banks, Voyager, Monster-Hunter, and Amoroso, to Oberea, Queen of Oteheite*:

> *One page from* HAWKESWORTH, *in the cool retreat,*
> *Fires the bright maid with more than mortal heat;*
> *She sinks at once into the lover's arms,*
> *Nor deems it vice to prostitute her charms;*
> *'I'll do', cries she, 'what Queens have done before';*
> *And sinks,* from principle, *a common whore.*

The literature of the South Seas was growing rapidly with these interminable satires, Hawkesworth, Parkinson and, in the future perhaps, publication of the botanical results of the *Endeavour* voyage.

While in Holland Banks took the opportunity to visit Pierre Lyonnet, a master printer who specialized in botanical engravings, and was very impressed with the work of one of his employees, Gerald Sibelius, who he would keep in mind for the plates of his own publication — the *Florilegium*. Botanical printing was popular at this time in continental Europe, where the world's leading publishers were concentrated, not only as a scientific record of plants but also as popular recreation. There was intense interest in the *Endeavour* plants from botanists all over Europe and particularly in Sweden, where Linnaeus was anxious to know what progress was being made in studying the material brought back. When he had heard of the likelihood of Banks and his favourite pupil, Solander, sailing away on the *Resolution* he feared that work on the important new plant discoveries in Tahiti, New Zealand and New Holland would be shelved. He wrote to John Ellis in late 1773 with some concern: 'Consider my friend, if these treasures are kept back what may happen to them. They may be devoured by vermin of all kinds. The house where they are lodged may be burnt. Those destined to describe them may die'. More than two years had passed since the expedition's return and it was not unreasonable for the great botanist to plead with his friend: 'By all that is great and good I entreat you, who know so well the value of science, to do all that in you lies for the publication of these new acquisitions that the learned world may not be deprived of them'.

As if to answer Linnaeus' plea, Kew and the *Florilegium* became Banks' main preoccupations. By the end of 1773 he had engaged a team of artists to make finished colour drawings based on the notes and outlines left by Parkinson. During the voyage he had produced 955 botanical drawings, of which 280 were finished colour renderings with botanical notes, and the rest—675 of them—quick sketches. From Tahiti onwards Sydney Parkinson was overwhelmed by work, partly because of the death of Alexander Buchan and also because of the increasing number of specimens collected in the Society Islands, New Zealand and on the east coast of New Holland. These London artists had a long task ahead of them, working from colour notes and referring to the original specimens lodged in Banks' herbarium. In the meantime, Banks had been seeking out good engravers to begin their work on the project as soon as possible. He sent specimen drawings to the Akademie der Wissenschaften in Berlin earlier in the year and by July was able to inspect the trial proofs pulled from the engraved plates. The Akademie was too expensive, however, and the prospect of supervising a large edition at such distance was too daunting. Banks discovered that the best London engravers were already engaged on projects that would take several years to complete but although they were unable to work for him, he studied their techniques and gauged the standard of work he could expect for his own project. The American scientist and politician Benjamin Franklin was interested in Banks' activities and noted in October 1773 that he was employing ten engravers on the project and kept a strict control on the quality of the work: 'He is very curious so as not to be quite satisfied in some cases with the expression given by either the graver, etching or mezzotint particularly where there is a woolliness or a multitude of small points on a leaf'. Unlike the artists who were completing the watercolour drawings from the later part of the *Endeavour* voyage, the engravers employed by Banks were not well-known for botanical work and a total of 18 of them would be used in the years ahead.

Now that he was spending more time at work and relishing the challenge, Joseph Banks was no longer a talking point in the capital, at least according to the man of letters, Horace Walpole, who was a close and rather precious observer of the social and political scene. Walpole informed Sir Horace Mann in a letter: 'Africa is, indeed, coming into fashion. There is just returned a Mr Bruce, who has lived three years in the court of Assyria, and breakfasted every morning with the maids of honour on live oxen. Otaheite and Mr Banks are quite forgotten'. That would soon change with the return of the *Adventure*, but in the meantime Banks was enjoying his period of relative obscurity, toying with the idea of colour for his *Florilegium*. This was a viable technique at the time, done either by applying the inks to an engraved plate and printing directly, or by later hand-colouring after a black impression had been struck.

This relatively quiet period did not last long. When the *Adventure* docked at Portsmouth in July 1774, it carried Omai, a native of Tahiti. Omai had requested to accompany the travellers, and Cook and Furneaux had agreed, since he would be useful as interpreter. Also, a return to Tahiti was planned, when they could offload him. As it happened, they did not go back and Omai achieved his secret ambition to visit their country. The *Adventure* had become separated from the *Resolution* after failing to meet a rendezvous at Queen Charlotte's Sound in New Zealand and made her way home independently.

Furneaux took his charge directly to the First Lord of the Admiralty in London, who immediately summoned Banks and Solander to come and greet this unexpected visitor from the South Seas in what they might remember of his native tongue. Omai recognized Banks immediately from the *Endeavour*'s stay at Matavai Bay and greeted him warmly. This set Bank's mind racing, remembering the plans he had laid for poor Tupaia's entry into British society. Here was an unsolicited replacement who could be adopted as the exotic pet he had wanted to acquire. Within three days of the confused native's arrival, Banks had Omai in his flowing robes presented to the King and Queen at Kew palace and suddenly the forgotten adventurer and his 'Indian from the South Seas' added a flutter of excitement to the jaded high society of the capital. The man who Cook had described as a 'dark, ugly, downright blackguard' was now viewed in the northern European context as tall, genteel, and well-made, in fact, the embodiment of the noble savage. Omai was a wonderful coup for Banks and the London newspapers competed with each other to report their adventures in the capital and all over Britain. Attempts were made to give Omai some basic education and he was watched carefully to assess the results of European influences on his untutored, unenlightened mind. The bluntest description of the Tahitian's presence came from a close friend of Banks, the Rev. Sir John Cullum, who estimated he was about 30 years old, rather tall and slender with a flat nose, thick lips and long black hair combining to make a disagreeable countenance. Omai was said to be possessed of a childish sense of fun and a curiosity about everything that was unfamiliar to him, particularly hail, ice and snow, which he called white rain. He had acquired some European manners from his long journey on the *Adventure* and after a short time in society accepted English food and drink and conformed to many of the accepted customs.

Banks received acclaim through the Tahitian's presence, but this reflected glory palled after several months and it became obvious there were less frivolous matters to attend to. Omai was put into lodgings not far from New Burlington Street where Banks could continue to keep an eye on him and at year's end a set of accounts detailing the native's expenses was prepared by Banks' book-keeper. It was revealed that 1774 had cost a total of £105.9.5d, a modest enough outlay for the amount of pleasure his patron had received in return.

Banks had been appointed to the council of the Royal Society and became closely associated with the affairs of its administration, attending nearly all meetings and dinners. There was also the demanding work at Kew, the supervision of his estates and the continuing production of the *Florilegium*. As the colour sketches were being completed, the exacting task of transforming the fine detail into delicate lines of copper engravings was being carried out, but very slowly, causing Banks to wonder if he had underestimated the time and effort needed to complete the edition. He was not prepared to compromize on quality, however, and it was inevitable that progress would be painstakingly slow. While this was going on, Omai shivered his way through the winter and dreamed of home.

Banks reached the age of 32 in February amid speculation in his circle about another distant journey of exploration. The Royal Society was keen to discover if a north-west passage across the top of America from the Pacific to the Atlantic existed. The Admiralty was interested in the proposal because of its obvious strategic and economic consequences, but had to hold back a decision about participating until Cook returned from the Pacific. The future of Omai was also considered at this time in official conversations and it was generally agreed he should be sent back to the Society Islands on the next British expedition headed for the Pacific, perhaps the one in search of the north-west passage. The *Gentleman's Magazine* fuelled discussions about Omai's destiny with two widely-read contributions: one of them predicting a particularly bleak future for the 'simple barbarian' who, they said, had been debased, and who after a course of 'improved debauchery' was to be sent back whence he came, full of the contagion of English vices, to revenge himself on his enemies and then die. A second article gave details of Bougainville's own noble savage, Aoutourou, who had generated the same sort of excitement for Parisian society as Omai in London. But Aoutourou did not live to resume his life in the Society Islands, having died of smallpox before departing from France. In spite of gloomy prognostications, the Tahitian seemed to be singularly uncorrupted by his time in England. There was talk about some capers with a well-known London transvestite, suggestions of amorous moments with the Duchess of Devonshire and speculation about marriage to an English woman who would accompany him back home, although few people saw anything remarkable about this type of gossip, which was an integral part of Georgian England's entertainment rather than informed comment. As they saw it, all it did was dispel the fashionable myth of the untainted savage living at one with nature. Dr Johnson was certainly no subscriber to this theory: 'The inhabitants of Otaheite and New Zealand are not in a state of pure nature; for it is plain they broke off from some other people. Had they grown out of the ground, you might have judged of a state of pure nature'.

At the height of the English summer of 1775 a report of Cook's imminent return from the Pacific was received in a letter despatched from the Cape of Good Hope in March. It documented some of the landfalls made by the *Resolution*, including Easter Island, the Marquesas, the New Hebrides, New Caledonia and the sighting of mountains of ice in the far southern seas, as well as the more familiar calls to New Zealand and the Society Islands. The elusive southern continent remained undiscovered and Cook had to make another apology for its non-appearance: 'If I have failed in discovering a continent it is because it does not exist in a navigable sea and not for the want of looking'. Daniel Solander in London wrote to Joseph Banks, who was with Omai on Lord Sandwich's yacht, giving a precis of the main information from the *Resolution*'s voyage, including the news that the botanists had collected some 260 new plants and 200 previously unknown animals. Banks must have felt far removed from all this as the touring party attended a dramatic performance in Plymouth, sitting through 'The worst acting we had seen for some years'. But this proved no great disappointment because, 'Many of us were of the opinion that plays should be either very well or very ill acted to be entertaining' and they went to bed well satisfied.

Soon after Cook's return the Admiralty confirmed that a new expedition would be sent to the South Seas and the *Resolution* was to carry Omai back home. Cook himself did not expect to command this journey and he was looking forward to a period of ease on land after many years of almost continuous sea travel and hardship. Banks was kept informed of these developments while he stayed in the country and Solander wrote to him with details of what the *Resolution* had brought back to England: there were three live Tahitian dogs, 'the ugliest and most stupid of the canine tribe', eagles and other birds, together with live springbok from the Cape for presentation to the queen. Cook continued to insist that he would not travel to the Pacific again, saying he was 'in hopes that I had put an end to all voyages of this kind to the Pacific Ocean, as we are sure that no southern continent exists there, unless so near the Pole that the coast cannot be navigated of ice and therefore not worth discovery'. He admitted, however, that a voyage would need to be taken soon to send Omai home. By the end of 1775 Banks' bookkeeper drew up the accounts for the Tahitian's expenses over the past 12 months and they totalled the rather daunting amount of £317.11.11½d. As the costs of the *Florilegium* were also mounting, Banks would not be sorry to see Omai's departure.

The Admiralty and the Royal Society decided to offer a prize of £20,000 to the first navigator to negotiate the northwest passage and there was also to be £5,000 reward for the first to sail within one degree of the North Pole. Cook was finally persuaded to change his mind and head the expedition on the *Resolution* with an accompanying vessel, another Whitby collier to be commanded by Charles Clerke from both the *Endeavour* and *Resolution* voyages. Before the departure however, it was decided that Omai should receive some religious teaching, which was initiated by Sir Harry Trelawney, to remedy what he considered were serious deficiences in the native's civilizing process while under the influence of Lord Sandwich, who lived openly with his mistress, and Joseph Banks, who had absolutely no reputation as a churchman. Sir Harry wrote to the Society for Promoting Christian Knowledge suggesting they should arrange for Omai to be instructed and ordained so that he could spread the word among his own people. Accordingly, the Tahitian received a month of tuition and was urged to learn the Ten Commandments, the Creed and the Lord's Prayer. This had little lasting effect because the language was almost impossible for him to memorize and he could not accept that adultery was a sin: 'Two wives—very good; three wives—very, very good', he was heard to comment. Preparations for this next expedition into the Pacific were prolonged because of the immense amount of special equipment and stores needed for a voyage both to the tropics and to the colder climates where the attempt would be made to find the north-west passage. Omai began a series of farewells, taking leave of the king and writing notes to many of his friends before departing from London in mid-June. 'Omiah to take leave of his good friends', his cards stated, 'He never forget England. He go on Sunday. God bless King George. He tell his people how kind English to him'. Omai had wanted to take two things back with him above all others: port wine and gunpowder, but Cook would not allow the powder, assuming 'some ambitious design'.

With his visitor gone at long last, having cost him £395.8.9d in 1776, including a tailor's bill of £86, Banks was able to enter a new and more settled phase of his life. The house which his sister shared with him in New Burlington Street had become too small for his many social activities and the storage of specimens, not to mention the constant stream of scientists eager to discuss the Pacific collections and the continuing work on the *Florilegium* plates. He bought larger premises at 32 Soho Square during the summer and moved in the following year. Another of Linnaeus' pupils, Jonas Dryander, came to work for him to catalogue the collection of plants, minerals, shells, insects, fish, birds, scientific instruments and all the artifacts shipped back on the *Endeavour*. This material was freely available for study to interested visitors and the house became a congenial centre of scientific discussion and philosophical conversation, in the best eighteenth-century tradition. The engravers engaged on the *Florilegium* worked with a pure line directly onto polished copper with the sharp engraving tool known as a burin. Each fine detail relied on the thickness and depth of a cut for accurate representation when printed. The deeper and wider incisions held more ink and registered darker; lighter areas came from shallower engraving. In this way every fold of a leaf, each contour of a stem and the subtlety of flowers could be depicted. Much of the exacting task needed to be done with magnifying glasses.

Although Banks was still young in years, his considerable experience, influence and fortune were leading him inexorably to the centre of the nation's scientific and intellectual life and he was finally being embraced by the influential society that his parents had wished for him so many years ago. In 1778 at the age of 35 he had already been a Fellow of the Royal Society for 12 years and a council member since 1774. When Sir John Pringle, the president, announced his retirement Banks was the obvious successor and was elected to the chair with an overwhelming majority. He took the honour very seriously and for the first time in his life had a clear direction. Britain was at the start of a scientific and technological revolution that would lead to her emergence as a major global power. Banks sensed that he was at the centre of these developments and was happy to expand his personal role as a benevolent autocrat: a part he played to perfection. His future was now clearly charted and the rich young blade was transformed into a pillar of the establishment. It was a convenient time to allow career and matrimony to merge and, a year after gaining the presidency of the Royal Society, Banks married Dorothea Hugessen, the daughter of a wealthy landowner, an heiress whose eventual endowment in Kent when linked to her husband's East Anglian estates, would make them very rich indeed. Unlike many of Banks' previous conquests, Dorothea was no beauty, 'handsome' was the most lavish praise that was applied to her looks, but the marriage was serene and successful, surviving the jibes that would be with them all their lives about his sexual affairs in the South Seas. Dorothea was happily compatible with Sarah Sophia Banks and the three were able to evolve a harmonious menage at Soho Square, while Banks' mother continued to live at Chelsea. Sarah Sophia supervised the running of the house, looked after her brother's affairs and was an avid collector of coins. A contemporary observer noted that her dress was usually of 'the old school' and her Barcelona quilled petticoat had a hole in either side for the convenience of rummaging in two immense pockets stuffed with books of all sizes. She was followed everywhere by a tall servant carrying a long cane. Sarah Sophia was a formidable woman who was entirely devoted to her brother, as experienced by a guest who happened to arrive early one evening while she was still supervising the preparations. Making small talk he observed, 'It is a fine day ma'am'. 'I know nothing about it', she replied, 'You must speak to my brother upon that subject when you are at dinner'.

Banks was considered the authority on the previously unknown territories he had visited in the Pacific and in 1779 was invited before a committee of the House of Commons that was enquiring into the state of the gaols in Britain and the question of transportation of felons. He spoke favourably of Botany Bay as a place for settlement, suggesting that its soil and climate would enable any occupants to become self-supporting. He thought there would be little probability of opposition from the natives, as during his own stay there, nine years before, he saw very few of them. The rest of the country seemed to be thinly peopled; those he had seen were naked, treacherous and armed with spears, and he thought them 'extremely cowardly'. The climate in the region of Botany Bay was comparable to Toulouse in the south of France, there were no beasts of prey and Banks had no doubt that sheep and cattle would thrive and increase on the pastures. The grass was long and luxuriant, there were plenty of fish, edible wild vegetables, particularly a type of spinach, together with water and timber. He was asked how a colony could best survive at the beginning of settlement and suggested they should arrive with a year's provisions and all the tools needed for tilling the soil and building houses. Cattle, sheep, hogs and poultry would need to be taken as well as grain and garden seeds. There should be ammunition for defence against hostile natives and, assuming that all went well, 'they might undoubtedly maintain themselves without any assistance from England'. Banks recommended sending a few hundred people at first and assured the committee that escape for the convicted felons

Portrait in pastels of Joseph Banks, by John Russell, 1778.

would be virtually impossible because Botany Bay was far removed from any other part of the globe inhabited by Europeans. Asked if Britain could expect any direct benefits from a colony at Botany Bay, he replied: 'If the people formed among themselves a civil government, they would necessarily increase, and find occasion for many European commodities; and it was not to be doubted that a large tract of land such as New Holland, which was larger than the whole of Europe, would furnish matter of advantageous return'. The committee made no specific recommendations about where to send felons but they were receptive to the idea of 'establishing a colony or colonies of young convicts in some distant part of the globe'. There the matter rested for the next few years, although others were also thinking about the suitability of New Holland for settlement. James Mario Matra would return to the scene with a proposal to send displaced American loyalists to New Holland and Banks advised the government on its official

146

plan for despatching a fleet of convicts to Botany Bay in 1787 to found a penal colony.

Progress on the *Florilegium* did not match the tempo of the rest of Banks' life, although at this stage about half the intended output was completed: 'About 550 plates are engraved', he was able to report to Linnaeus, 'and I think if circumstances as yet unexpected do not oblige me to cut it short, it will double that number'. By now there was doubt in many minds about Banks ever publishing an image or even a word from the voyage. He explained the delay in reply to a letter from Edward Hasted: 'Botany has been my favourite science since my childhood; and the reason I have not published the account of my travels is that the first, from want of time necessarily brought on by the many preparations for my second voyage, was entrusted to Dr Hawkesworth; and since that I have been engaged in a botanical work which I hope soon to publish, as I have now near 770 folio plates prepared: it is to give an account of all the new plants in my voyage round the world, somewhat above 800'. In 1781 a baronetcy was created for him and he chose as his personal symbol a lizard, which, he said, was an animal endowed by nature with an instinctive love of mankind. He took it for his seal: 'as a perpetual remembrance that a man is never so well employed, as when he is labouring for the advantage of the public, without the expectation, the hope or even a wish to derive advantage of any kind from the result of his exertions'.

In May 1789 Daniel Solander suffered a stroke and died after lingering, paralysed, for a few days. He was 49. For Banks, who had been at his bedside, it was the greatest sadness he had ever experienced: 'This all too early loss of a friend whom I loved in my riper years and whom I shall always miss means that I wish to be allowed to draw a curtain over his

Left: Portrait of Sarah Sophia Banks by John Russell, 1779.

Right: Portrait of Dorothea Banks (*née* Hugessen) by John Russell, 1779.

Banks' study at Soho Square.

decease, as soon as I have ceased to speak of it. I can never think of it without feeling a mortal pang, at which mankind shudders. But if modesty, justice, moderation, and both natural and acquired skill can claim a place in a better world, nothing but a lack of equal merit on my part can prevent us from meeting again'. The funeral and burial at London's Swedish church was attended by many of Solander's countrymen.

Although Solander had completed his manuscript volumes containing the descriptive text for the whole *Endeavour* voyage, his loss meant a severe jolt to the progress of the *Florilegium*. A couple of years later, with undue optimism, Banks noted that the huge enterprise was nearing its completion and the main credit for the work would be shared equally between the two of them, with Solander's name appearing on the title page next to Banks'. 'While he was alive', Banks pointed out, 'there was hardly a passage composed on which he was not represented. Since all the descriptions were made when the plants were fresh, nothing remains to be done, except to fully work out the drawings still not finished, and to record the synonyms which we did not have with us or which have come out since'. Banks was confident towards the end of 1784 that so little was left to be done that it could be completed in two months, 'if only the engravers can come to put the finishing touches to it'. He did not explain why it was necessary to complete all the copper plates before starting publication as he had already

148

indicated an intention of releasing the work in several sections and by this time there was more than enough material to make a start on the printing. It was obvious that Banks, always the gentleman amateur, never a real scientist, had simply lost interest because he had no need to publish. After Solander's death, Dryander moved into 32 Soho Square to become Banks' second botany librarian, but the atmosphere was now changed, as the Swedish court pastor noted in a letter home about him: 'Comparisons are always bad, but I cannot refrain from saying it, he is not Dr Solander, in whom the Swedes in London have lost more than I can describe in words'.

An intense involvement with the Royal Society and Kew further lessened Banks' attention to the project and many other official and personal commitments could only lead to more delays. By then the world as seen from Britain was becoming an unsettled place and his own priorities pushed the *Florilegium* into the background. As Hamlet soliloquizes: 'It is an unweeded garden that turns to seed', so the great publishing project languished. Had Buchan and Parkinson survived the *Endeavour* voyage the finished watercolours would have been completed much earlier; had Solander lived and spent less time on his social rounds he could have supervised the completion of the edition; and if Banks had not been such a busy man with a multitude of responsibilities and interests, the engravers' work might have been speedier. There was still lingering hope in scientific and publishing circles that the *Endeavour* material might appear soon because its collector and patron was barely 40 years of age and appeared to be in splendid health, although it was rumoured he suffered increasingly from gout.

Fame had been an early spur to the *Florilegium*'s publication, but after 13 years Joseph Banks was a greater celebrity than he could ever have imagined and did not need to make his name. There were also economic reasons for allowing the great project to stagnate. The uncertain financial climate of the 1780s, preceding the Napoleonic Wars, caused a sharp drop in rural incomes. This severely affected the Banks' estates in Lincolnshire and Kent, already

Men of Science living in 1807–8. Designed by the engraver William Walker. Banks is standing behind the fourth seated figure from the left.

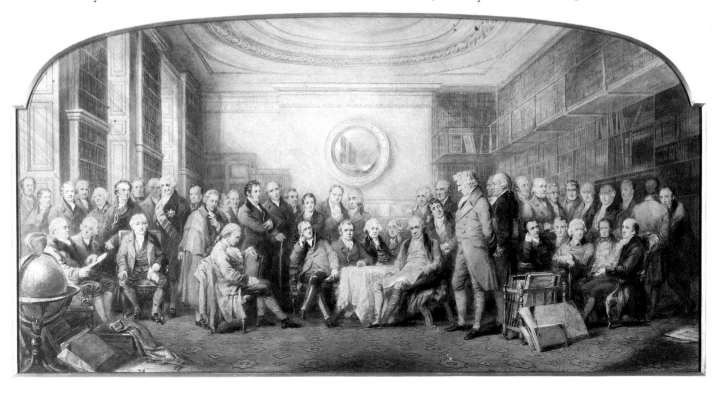

Pub. July 4th 1795 by H. Humphrey No. 37
New Bond Street

disadvantaged by the depression in the wool trade after the American war of Independence. Any hopes of marketing a publication of the *Endeavour* flora in Europe or America were dashed. Towards the end of the century there were riots in Lincolnshire directly connected with the state of war with post-revolutionary France. All hired labourers, including those in Banks' employ, were required to join the militia for call-up at a moment's notice in the event of invasion. Many rebelled against this compulsion to fight for what they saw as protection of the rich. Banks suggested clemency for the convicted rioters rather than death by hanging, followed perhaps by transportation to Botany Bay. By now, any inclination to produce the *Florilegium* had passed.

Banks still had two decades of activity ahead of him. He maintained an intense interest in the development of the new colonies in Australia and assisted in founding their wool industry. He also continued his duties at the Royal Society and Kew. He was honoured by almost every learned society in Europe so that his rule among men of science was as absolute as Dr Johnson's had been in the world of literature. Increasingly plagued by gout, he made his last journey from London to Revesby in 1817. In June 1820, a year after the death of Sarah Sophia, Sir Joseph died at Spring Grove, his estate west of London. He was 77 years old and left no heirs.

The local parish church where he was buried at Heston in Middlesex has since been rebuilt and stands behind its leafy chestnut facade among the fumes of a busy thoroughfare. The constant noise of traffic mingles with the sound of jets on their final approach to London's Heathrow Airport. A modest tablet is the only indication that Banks' body lies somewhere below the nave. His will requested internment in the most private manner, entreating his surviving relatives to spare themselves the affliction of attending the ceremony and asking that they erect no monument to his memory. The wish was granted, and it would be curiously appropriate that his lasting memorial became paper rather than stone.

Spring Grove at Isleworth, Banks' final home. Watercolour by John Preston Neale, 1816.

Opposite: Caricature of Joseph Banks by James Gillray. Handcoloured engraving, 1795.

151

10

The Long Wait

1820–1979

If you can paint one leaf you can paint the world.

John Ruskin (1819–1900)

NEGOTIATIONS with the British Museum over the extensive collections offered them under the terms of Banks' will were completed in 1827 after four years of discussions. This material included manuscripts, drawings, engraved copper plates and the contents of Banks' herbarium and library. Banks' curator/librarian, Robert Brown, served as the official keeper of the Banksian Botanical Collection at the museum for the next seven years, until the trustees decided to place all the botanical collections under his care, which he supervised until his death in 1858. Nothing happened about printing the *Florilegium* plates during this time; the collection in its paper wrappers remained on the shelves in Bloomsbury and gathered dust. In fact, publication was not seriously considered until 70 years after Banks' death when William Carruthers, the Keeper of Botany, urged the trustees to consider an edition from the engraved plates as part of a series illustrating the plants collected on Cook's first and second voyages. By now the natural history collections had been transferred to a splendid new building in South Kensington, to be known officially as the British Museum (Natural History). Detailed reports were submitted to the trustees but they were unable to reach a decision.

The delay in publishing the botanical discoveries from the *Endeavour* voyage meant that others were first with descriptions of the flora of the South Seas. Robert Brown in his *Prodomus Florae Novae Hollandiae* set out to describe all the species of plants he had collected while travelling as botanist on the *Investigator*, under the command of Matthew Flinders, between 1801 and 1803, together with material from Banks, Solander and others. His first volume, without illustrations, had appeared in 1810, but the project did not proceed because of poor sales. It seemed there was not a large enough market for a purely scientific work such as this. In nineteenth century Britain there was a sizeable number of educated people who were unashamed flower lovers but, like the writer John Ruskin, resented every botanical term being turned into Latin or Greek. Ruskin wrote a book about flowers, which were one of the great passions of his life, and as a poet and painter he saw himself 'at war with the botanists'. In the 1860s, while the British Museum trustees were still considering publication, Bentham's *Flora Australiansis* was published with such a wide selection of plants that it needed to make reference to only a handful of the *Endeavour* discoveries. When Drake del Castillo produced his volume of Tahiti flora in 1893 most of the novelty surrounding Banks' and Solander's work had evaporated; the finely-engraved plates ceased to hold much scientific significance, being now mainly of historical interest.

The next stirring of curiosity in the *Endeavour* material came from outside the museum in June 1893, when the celebrated naturalist, Sir Joseph Hooker, sought permission to print a transcript of Banks' journal from the voyage. His request was deferred but the indefatigable 76-year old, who was a leading disciple of Darwinism and had been an intrepid explorer in his younger days, was persistent and permission to publish was finally granted. The trustees had good reason to wonder if they had acted wisely when his 557-page tome appeared in 1896. Hooker's version of Banks' journal contained many 'curtailments' of the original text and with Victorian thoroughness and the liberal application of red ink, he had deleted, re-written and censored, until fully a third of the original material disappeared. Late in 1894 Sir James Hector, a Fellow of the Royal Society and a former head of the Geographical Survey of New Zealand, who was then director of the New Zealand Institute, asked for access to the plates of his country's plants and manuscript entries for a new flora being prepared by Thomas Kirk. It was planned that the proof sets of New Zealand engravings should be used in reduced form as illustrations. The trustees were in agreement, but Kirk died in 1897 before he could complete his work and the book was never published.

George Murray succeeded Carruthers as the museum's Keeper of Botany in 1895 and, after the failure of the New Zealand venture, he submitted a proposal for publication of the flora from Cook's first and second voyages. The trustees came back to him with visions grander than his: they wanted 500 copies and requested costings from foreign companies to find out if printing abroad would be cheaper than in Britain. Murray worked quickly; within two months, he managed to compile a proposal that was as economic and practical as possible, suggesting use of the cheaper method of photolithography for an edition limited to 300 sets. He considered the estimate from the British firm of Hazell, Watson and Viney the most acceptable, totalling £1,273.15.0 for plates and text. The trustees agreed to proceed on that basis, once the financial estimates for the following year were approved, and so, nearly ten years after Carruthers' original proposal, and 80 years after Banks' death, the first part of the work was published under direction of James Britten in April 1900 as *Illustrations of Australian Plants*. Sets of 100 monochrome prints were issued that year, produced from lithographic stones prepared from the original copper engravings, accompanied by Solander's Latin descriptions edited and updated by Britten in accordance with the classification procedures then current. The second part, comprising 143 prints, appeared in 1901 to coincide with the federation of the Australian colonies into a commonwealth, and the final section of 75 was completed during 1905, making a total of 318 prints.

The Australian material comprised less than half the number of engravings originating from the *Endeavour* voyage and their production alone had

taken five years for an outlay of £837. It would cost at least that amount again to complete the entire complement and from the point of view of science it was probably not worth the effort, because the great majority of the subjects remaining unpublished were already well-known to botanists. Accordingly, the trustees called a halt to the printing after the completion of the Australian material as neither the expense nor the continuing drain on the time of the museum staff could be justified. The Empire was now in a position to strike back, and J. H. Maiden, the director of the Royal Botanic Gardens in Sydney and a Fellow of the Royal Society, was highly critical of the disruption to a complete Banks edition. In reply to his disapproval, Britten said: 'I am probably the last person to deprecate the value of a work in which I have taken the greatest interest, but it seems to me on many grounds it may be doubted whether the actual scientific value of the book, apart from its historical interest, is equivalent to the expense necessary to its production; and this reason has so far weighed with the trustees that the remainder of the plates of the plants of Cook's voyage will not be proceeded with'.

In Sydney during the first decade of the twentieth century the committee of the Banks Memorial Fund wanted the memory of the man they regarded as Australia's benefactor to be enshrined in some enduring recognition, and they suggested a statue for a prominent site in the city or perhaps at Botany Bay. One-hundred-and-twenty years after the founding of the first European colony in this land on the shores of Port Jackson, Australia was in search of a founding father. J.H. Maiden, who was also the Government Botanist of New South Wales, agreed to become honorary secretary of the fund. He wrote a book, *Sir Joseph Banks, the Father of Australia*, which was published in Sydney in 1909 by the Government Printer and also in London. In an introduction he stated his motive in writing the biography: 'We know Sir Joseph Banks chiefly as a great name, but our knowledge of him is vague; we cannot be expected to contribute to a memorial to him without learning more about him'. Maiden received official backing and his book was printed at public expense on the understanding that there would be no free copies and all proceeds would augment the memorial fund. But the fund failed to flourish enough for a fine statue because James Cook was known as the nation's hero and the name of Joseph Banks slipped into desuetude. Not even a university scholarship, proposed in his name, eventuated. The memorial committee ceased to exist in 1937 when Sir Daniel Levy died, its last surviving member. At that stage the fund totalled £1,089.15.9 and in 1943 the New South Wales parliament passed the Sir Joseph Banks Memorial Fund Act, establishing a trust to 'consider how the fund may be utilized for the purpose of providing a suitable and fitting memorial to perpetuate the memory and services of Sir Joseph Banks'. Two years later a further act was passed, repealing the former one, which vested the fund in the trustees of the Public Library of New South Wales, 'to apply the same in or towards the cost of editing, publishing and distributing the Banks' papers in a manner and form suitable and fitting to the memory of Sir Joseph Banks'. The fund was transferred to the library trustees in March 1946 and then totalled £3,941.14.3. After a search for a suitable editor, Dr J.C. Beaglehole of Wellington, New Zealand, agreed to produce an edition of Banks' *Endeavour* journal and the two-volume work was first published in 1962. At long last, there was an accurate record of the epic journey in Banks' own words, together with scholarly annotations.

Portrait of Joseph Banks as President of the Royal Society, by Thomas Phillips, 1816.

In the meantime, the story of Banks' plates in Britain was one of total neglect. From a scientific point of view the images were now only mildly interesting, not important enough to make publishing imperative, and their inherent aesthetic appeal seemed to interest nobody. The museum in South Kensington continued to be a centre of study for scholars from all over the world and the plates, together with the black impressions struck from them, had always been available for reference on request. The engravers were known to have taken about three sets in the eighteenth century and proofs of particular plates had been sent by Banks to other botanists for their private use. A group of 28 prints was recorded as forming a folio volume in the Berlin Library. Exactly how many plates there were was a matter of conjecture because sorting through a ton of copper to find out was too daunting a task. About 750 had been engraved from completed watercolour drawings by Banks' staff from 1772 to 1784. The catalogue of drawings and a supplementary list of plates in the museum's archives suggested that 753 copper plates were selected for engraving. As 738 plates survive to the present day it seems that only 15 disappeared in the interval of 200 years. Being kept within the museum was the best security, although the collection was greatly endangered in September 1940, when the Botany Department was severely damaged by fire one night in a German raid. The next morning revealed a scene of devastation with dried specimens, precious books and catalogues scattered everywhere amid the dusty debris. Museum staff worked hard to save whatever they could from the rubble and the Banks plates were piled onto tables and left there in stacks for a considerable time before being placed in cupboards outside the Botany Library for the best part of a decade.

It was not until early 1963, when Dr William Stearn of the Botany Department had a few of the original plates proofed at the Royal College of Art to check the quality of the engraving, that any serious thought was given to another edition. Close examination by the printmaking department indicated that they were in much better condition than could be expected after nearly 200 years and it was decided to select a number of them for production on aesthetic rather than scientific grounds. Later a final choice of 30 would be determined by additional botanical or historical interest. Some people had questioned the quality of the plates and their relevance to twentieth century printing. Wilfrid Blunt in his book *The Art of Botanical Illustration*, published in 1950, thought the Banks engravings 'disagreeably mechanical', although he softened his opinion considerably when invited to write a preface to this new edition. The project was given the inappropriate title of *Captain Cook's Florilegium* and took ten years of extremely slow work before appearing in 1973 from the Lion and Unicorn Press, the publishing imprint of the Royal College, in a limited edition of 100 numbered copies in black ink. Cook's association with the botanical aspects of the *Endeavour* voyage, except as Banks' shipdriver, were rather tenuous and on several occasions he had been cynical of the gentlemen's flower-gathering activities. The captain's name in the title of this new edition could only be a marketing ploy to capitalize on the more famous personality of the voyage. In Banks' own time the term 'florilegium', literally 'a binding of flowers', was taken to mean a picture book of garden flowers compiled for horticulturalists rather than botanists. The edition was fully subscribed and found its way to public libraries and collectors in places as far afield as Los Angeles, Toronto, Sydney, Washington, Vienna and Brisbane. Later, a further deluxe set was produced incorporating three specimens of Australian plants encased in clear acrylic plastic — *Banksia serrata*, *Banksia integrifolia* and *Xylomelum pyriform*, gathered at Botany Bay in December 1973. This edition was bound in Nigerian goatskin with endpapers of pure silk embellished with 22-carat gold leaf. These much-delayed publications received considerable publicity, particularly for Captain Cook, because the bicentenary of the *Endeavour* voyage was still fresh in many people's minds. They also proved to be an outstanding investment for the original subscribers. The attention helped to re-kindle an interest in the Banks engravings just when it seemed as if the flames of his and Solander's endeavours were about to die. During the printing of *Captain Cook's Florilegium* five of the copper plates were lost, leaving a total of 738 survivors.

It was while the Banks plates were at the Royal College of Art during the early sixties that Joe Studholme, of Editions Alecto, first became aware of them. He was working closely with the college at the time on various projects and was able to see the results of the proofing as tests were made on the historic engravings. The enthusiastic Alecto partners discussed the project among themselves and thought what a triumph it would be to issue all the Banks engravings. With this in mind, an approach was made to the museum for the newly-incorporated Editions Alecto to print and publish the plates and, as Studholme says, they 'quite rightly got shown the door'. Ten years later, when *Captain Cook's Florilegium* was completed, Alecto's reputation had become internationally established as a leading publisher of contemporary prints and multiples with an enviable catalogue of artists, like David Hockney, Eduardo Paolozzi, Henry Moore and Jean Dubuffet. They also had an impressive catalogue of historical subjects, with editions of William Daniell, Hans Holbein and a portentous Dante's *Inferno*. The origin of the *Florilegium* project came peripherally from John Hawkins in Australia, a Sydney antique dealer with a great interest in Australian history and an expert knowledge of silver. He suggested that his friend Joe Studholme, now managing director of the company, visit the Natural History Museum in South Kensington to look at the Watling collection, thinking that the depictions of early Sydney might make a good subject for a limited edition of reproductions. Thomas Watling was a coach painter from Dumfries in Scotland who was charged in 1788 with forging banknotes and sentenced to transportation to New South Wales for 14 years. In 1792, at the age of 30, he became the first professional artist to arrive in Australia and soon found his talents in demand for recording the gradual development of Sydney town. Much of Watling's work went back to England and was lodged in the Natural History Museum. It was this material that Hawkins thought might interest Alecto. Joe Studholme appreciated the suggestion but declined to do anything about it: 'Thanks, but no. We're not in the business of making reproductions'. The visit to the museum, however, was not unfruitful. A person who was working as a salesman with Alecto at the time, Nigel Frith, was a neighbour of one of the Botany Department's curators, Dr Chris Humphries. They talked about Alecto's interest in historical printing and Humphries wondered if they might consider the Banks plates, since they had remained unpublished in a representative edition. 'This rang an enormously large bell' for Joe Studholme because of his interest in the same material more than a decade before. Visits to the museum led to discussions about the feasability of printing and publishing the material followed by more formal approaches to M.J. Rowlands, Head of Library Services, and Robert Cross, Head of Publications.

So Banks' *Florilegium* began to take shape through a fortunate set of circumstances. All previous attempts to print large editions of the plates had seen the trustees of the Natural History Museum take an extremely cautious line and there was little reason to think that their inherent conservatism had changed dramatically since William Carruthers spent nearly ten years getting his proposals accepted, and he had the advantage of being an insider. But now, nearly a century later, the omens were auspicious for another assault on this Parnassus of publishing under the favourable patronage of the Greek Furies, one of whom had given the company its name. The Alecto people were displaced persons because of a disastrous fire that had reduced their etching studios to a clutter of charred machinery, while printing an edition of Dante's *Inferno*. It was no time for formal negotiations of every last sub-clause in a complicated contract; there were too many imponderables for both the museum and the would-be publisher. The fire, which could have destroyed a lesser determination, never actually closed Alecto and it had the effect of recalling earlier barnstorming days, when they had had to sell prints to any market for survival. They moved up the road to Kensington High Street, took temporary premises below Régine's night-club and announced a fire sale, which turned out to be more successful than anyone could have imagined, selling quite a lot of stock that previously had been impossible to move. Alecto then took over the local greengrocer's shop, which was closing down, at Kensington Court Place, and work continued as the original premises at Kelso Place were cleaned up and restoration and rebuilding began. Before this, in September 1978, Joe Studholme had written to A.P. Coleman, Deputy Director of the British Museum (Natural History), suggesting a joint publication between them and Alecto Historical Editions of perhaps 300 plates printed in black in a proposed edition size of 150. The possibility of hand-colouring some plates for a small separate edition was also mooted.

The company's printers had always been encouraged to teach; as Joe Studholme says: 'It kept them sane in an otherwise boringly repetitive task and put them in touch with what was going on in the art schools and colleges'. Charles Newington lectured at the Winchester School of Art in Hampshire, where one of the students was particularly outstanding because of his enthusiasm for printing techniques and an obsessive urge to experiment with them. He was a slight, willowy young man with a shock of fair hair, named Edward Egerton-Williams. After his studies were completed he went to work for one of Alecto's subsidiary printing companies, Tisiphone, named after another of the Greek Furies, and was engaged on William Daniell's *Voyage round Great Britain* plates. The young man was a perfectionist, always seeking new ways of producing better prints. He left Winchester in July 1978 with an ambition to set up his own studio but was realistic enough to know it was necessary to learn the business first and establish a reputation. He joined Alecto in September of that year at the age of 23 and remembers: 'In those days I used to work for Alecto during the day and then do quite a number of nights as a freelance printer'. His commitment to all facets of printing saw Egerton-Williams haunting the etching studios days, nights and weekends. At this time the Natural History museum allowed Alecto to have a couple of the Banks copperplates for making proofs to assess the possibilities of proceeding with the project. They were handed over to Egerton-Williams to work on with the warning that they were going to be difficult but he should see what could be done about getting reasonable prints from them. He moved quickly, striking the first prints from the plates that night before going out to dinner. Joe Studholme looked at them the next morning and was very impressed with the results. 'Are they a problem?', he asked. 'No, not at all, I just dashed those off very quickly'. A few more of the precious plates were trial-printed as the next step in a feasibility study, still with the general idea of producing a black edition from perhaps a third of the available originals. Egerton-Williams was working largely on his own in another studio, because the fire had meant Alecto's intaglio lithographic and screen printing studios had to be relocated in other parts of London. Having an extremely low threshold of boredom, he balked at the thought of turning out hundreds of separate monochrome prints: 'So I mixed up a lot of inks trying to make them pretty, and produced some of them in colour. They were not botanically accurate, far from it. We had no reference to the original finished watercolours, and so I started to make nice pink flowers with green leaves and brown stems, irrespective of their real tonings'. He was evolving his own techniques of colour printing by instinct and testing the possibilities of the medium: 'I knew nothing about botany, but the results looked fantastic'. *Banks' Florilegium* moved another step nearer realization during the spring and summer of 1979 with continuing experiments to achieve perfect colour proofs. The idea of hand-colouring monochrome prints was abandoned as being too slow and because watercolour paints tended to obliterate fine details in the engraving.

Master printer, Edward Egerton-Williams.

It was important now to study the finished watercolours in the museum, since they had provided the basis of the engravings. Joe Studholme deduced from their high degree of finish that Banks must have been considering publishing in colour; it was known he had been in touch with the French colour printers. Banks' own engravers had been instructed to render the subtle nuances of texture in order to give an accurate representation of the plants' structures, but he must have thought about the application of colour to heighten both the botanical fidelity of the record and also, perhaps, to give an added aesthetic to the subject. Nothing more specific was known about this, since there was no documented information about his precise intentions. Studholme admits to hopes of doing a colour edition, and these were heightened when Egerton-Williams' experiments indicated it was possible to produce good colour prints from the old plates. The first of these proofs were taken to Paul Hulton of the Department of Prints and Drawings at the British Museum for his opinion. He had just organized an exhibition called *Flowers in Art* and his reaction was immediate and encouraging: 'If Banks were looking at printing them now', he said, 'this is the way he'd have them done'. Without really thinking about the full implications of time and money Joe Studholme decided to print 'the whole lot' and a memorandum of agreement for publication of *Banks' Florilegium* was signed on 6 October 1979 by Robert Cross for the museum and Joe Studholme for Alecto.

A degree of reckless confidence was needed for what would be a lengthy, labour-intensive and, therefore, expensive project. Interest among potential collectors needed to be assessed so that orders for the huge number of prints might be received and the project started. The United States and Australia were considered prime markets because nearly all the material originated in the Pacific and there was a growing interest in Australian history in both countries. In late October 1979 Studholme and his wife Rachel boarded a plane for New York with a portfolio of proofs and a list of librarians and private collectors who might be interested in subscribing. During the six-week trip they also visited Melbourne, Canberra, Sydney, Brisbane and Perth and, as anticipated, there was considerable interest from public libraries, although none was able to give an immediate commitment because of the cost and scale of the work, which embraced 738 separate prints and would involve a production period of at least seven years. It needed careful consideration by trustees and boards of governors before commitments could be made. The best prospect of early sales came from collectors, who would see the work as an outstanding investment. Many of them gave the *Florilegium* proposal an enthusiastic welcome.

Another person who was instrumental in making the project happen was the Sydney publisher, businessman and television station owner, Kerry Packer. Packer was well-known in Britain and at home for changing the character of professional cricket through his World Series Cricket, making it a vital, crowd-pleasing sport; he was known only to a few as an avid collector of antique silver. He also owned a good art collection, including a number of outstanding Dobells inherited from his father, Sir Frank Packer, the newspaper and magazine publisher, who had been the first Australian challenger for the America's Cup. Kerry Packer retained John Hawkins as his dealer for silver and he arranged a meeting with Joe Studholme to discuss the *Florilegium*. The result was immediate and surprising: 'I thought he might take one of the 40 sets we had earmarked for Australia', Studholme remembers, 'but was amazed when he brusquely said he would take the whole lot'. This was the needed promise of enough financial backing for the entire project and an agreement was made to underwrite the allocation set aside for Australia and New Zealand with Alecto retaining responsibility for the sales and distribution. Kerry Packer paid for the two sets he wanted for personal acquisition, knowing that it would be several years before he received the goods in full, thus becoming the first subscriber to commit money to the project.

The size of the edition had evolved as haphazardly as the project itself. Alecto knew it would need to be limited to make it both financially and logistically practical. There could have been 125, or 150 sets, but the greater the edition, the longer it would take to complete. The number was settled at the round figure of 100, and there were to be ten extra sets *hors commerce*, which simply means in modern terms outside the main edition. They would bear the numbering 1/X to X/X, with each sheet of the entire edition stamped with the publishers' marks. If the project had started in Banks' time it would have remained unnumbered and printed to demand. The idea of fine arts limited printing is a twentieth century convention, as Joe Studholme puts it, 'a bogus exercise really. We do it simply because there has now been a long tradition for limited editions, it is, in fact, a marketing ploy: we need to reassure people that there are only going to be a certain number of numbered sets'. Exclusivity in this unusual project was further assured by an agreement reached between Alecto and the Natural History museum that no more impressions would be taken from the plates for at least fifty years after the completion of publication. 'It's really just to keep up the value of the product', Studholme explains. 'You have to do it because of the tremendous investment involved'.

The financing of the *Florilegium* was, in the words of Alecto's managing director, 'awfully relaxed'. Kerry Packer's agreement to take a large part of the edition was the encouragement needed to start the printing, but Alecto soon realized it was not such a good idea for the flexible distribution of the sets. It was always intended that the great work should be placed as widely as possible, particularly among libraries and universities in those countries where the original collections had been made. An emphasis on the botanical nature of the production was one of the key factors in gaining the museum's agreement to allow their Banks material to be published by Alecto. It was unlikely that very many sets would find their way either to Chile or Argentina, depending on who currently laid claim to Tierra del Fuego, but Britain, Australia, New Zealand and France — because of the Tahiti connection — and possibly Brazil and Indonesia might very well become subscribers. In addition, the traditional market in the United States for fine arts printing was vitally important for subscriptions, particularly if Studholme's 'always print less' dictum held true. Alecto realized that under their agreement they would act in Australia and New Zealand as agents for Kerry Packer in seeking subscribers for about 38 sets allocated to the region but not yet taken up. It was thought this might inhibit the flexibility of their world-wide marketing and the deal was re-negotiated, leaving Kerry Packer with the prized first and second sets of the *Florilegium*, together with a third subscription for his company, giving him the additional personal kudos of being one of the people responsible for initiating the largest ever direct printing project in the fine arts. Alecto took back the rights to the Australian and New Zealand allocation confident that, as Joe Studholme put it, 'This is going to be a runner.'

Fate now decreed that the paths of two Josephs, Banks and Studholme, both from Eton and Oxford, should cross after 200 years. Alecto was to become the vehicle whereby this great botanical work would be published in its entirety, in full colour, for the first time.

11
Realization
1980

All art is but imitation of nature.

Seneca (?−AD 65)

THE next consideration was to find a studio for the printing because Alecto had no working premises of its own. Edward Egerton-Williams had nursed an ambition to own an etching studio and now seemed like the right time. He was appointed master printer of the project, signing an agreement with Alecto to provide all production services. This set the two principals, the tall, willowy Guardsman-like figure of Joe Studholme and the much shorter, slighter, master printer tramping the streets of London in search of suitable working space. They finally found it in a maze of back streets opposite a railway viaduct on one of the approaches to Liverpool Street Station in the shabby East End. Before a start could be made on the first plates the studio had to be set up. Edward did this with the help of a bank loan and some personal money, having the valuable assurance of several years' work ahead of him on the edition. The derelict but solid four-storey warehouse in Appold Street chosen for the premises looked umpromising, but it had one overwhelming advantage: natural light. Large windows stretched across two sides with a southerly and easterly aspect, a great bonus because good illumination would be the basis of the quality control so essential to the success of the project.

The Egerton-Williams Studio, as its shiny brass plate would soon read, was on the third floor of the building. The initial four or five workers included Mike Barratt, later to run the proofing, and Paul Brayson, who left eventually to become a successful portrait painter. Over a period of three months they cleaned out the place, which was filthy, cleared the former fittings and fixtures and installed presses, benches and other equipment. The white walls were decorated with contemporary prints, the floor was sanded and oiled, they painted bright primary red on the supporting iron pillars and brought in a profusion of indoor plants that thrived in the natural brightness. It was a dramatic transformation. Work began with a loan of £10,000, but just three weeks later the overdraft had soared to £27,000 and Edward found himself writing cheques for supplies and equipment he had

163

Exterior of the Egerton-Williams
Studio in the East End of London
where the *Florilegium* is being printed.

no hope of covering in the short term. He was realizing his ideal, a 'big
toy', as he later described it, although fiscally it seemed a reckless thing to
admit: 'We don't get on very well with our bank'. There were numerous
warnings at first: 'Don't you realize that nobody ever makes a success from
such publishing? — and you're doing it in the aftermath of a fire!' 'You have
this lad as a master printer who's frightfully cantankerous, has difficulty in
getting on with people and looks as if a puff of wind would blow him over:
what are you up to?' These comments were ignored and, with complete
confidence in Egerton-Williams' ability to deliver the goods, Studholme
watched the progress.

The priorities were to get the project moving by devising the best methods
of producing prints in colour by techniques that were unfamiliar to modern
printers. Egerton-Williams had never heard of anybody using a process of
rubbing inks into an etched plate, although it seemed to be a perfectly
reasonable way of going about it. 'There was a lot of bullshit about how
things should be done', he says, 'or more exactly how they *might* have been
done. To me this was just a job that happened to come up and needed to be
solved'. They began on the largest section of the *Florilegium*, with prints
from the east coast of Australia, 337 plates in all. It was going to be a long
haul of nearly three years, longer than the *Endeavour* voyage itself, before a
start was made on Brazil. Later would come Java, Madeira, New Zealand,
the Society Islands and Tierra del Fuego. The first impressions produced in
the new studio justified Joe Studholme's faith in Egerton-Williams' ability
to produce colour of the finest quality.

In August a prospectus was published for the edition and two months
later a launching party was held at the now refurbished 27 Kelso Place with

representatives of all the countries where Banks and Solander collected. On 18 November 1980 the formal announcement of the forthcoming publication of *Banks' Florilegium* was made during a reception at the Natural History Museum. Subscribers now had to be sought for the expensive edition. The initial scheme was to bind the prints in books, but then it was realized they would probably be used more frequently as loose sheets for framing or study. According to Joe Studholme, they started with the idea of 'gold leaf and lots of leather. Then we thought that was nonsense because the plates were so good in their own right they needed to be presented in the most restrained way'. That is how Alecto arrived at the idea of placing them in mounts of the same quality paper.

Above: Joe Studholme, managing director of Alecto Historical Editions in the *Florilegium* studio.

Below: Edward Egerton-Williams discusses a Parkinson watercolour with one of his printers at the British Museum (Natural History).

The methods adopted in the studio appeared deceptively simple to casual observers, but behind the apparently smooth operation lay months of trial and error and more than a little frustration. As the precious plates began to arrive from the museum it was clear many of them had remained untouched since long before Banks' death. At some stage their wrappers had become damp and the acid in the paper transferred to some of the surfaces etching itself in random patterns on the copper. A few of the Australian subjects had such bad foul-biting, as it is known, that the engraved images were hardly discernible. Another problem was that when the original engravers had taken their proofs, they did not bother to clean the plates afterwards, probably because they expected them to be printed fairly soon, thus leaving delicate lines badly clogged with dried ink. This ink was originally made with a base of linseed oil and there was no solvent readily available to break down its rock-hardness. The images needed to be stripped of the residue before repairs could be carried out to the acid etching and scratches. Unless every single line was absolutely clean, it was impossible to obtain a fine impression. Egerton-Williams was able to find a painting restorer who suggested several chemicals that might do the job, but the process, alternating the solvents with hot soapy water, proved to be excessively slow. When a plate was finally stripped to the clean copper, it was put in the press and a black print pulled to check its condition. Repairs could then begin using scrapers to take the metal below the level of imperfections, after which the surface was polished with a burnisher. It was soon discovered that normal tools were too soft for this particular work and had to be chrome-plated for extra hardness. At this stage, most of the plates would be ready for protective coating before printing, but close examination sometimes revealed a flaw. The 200-year-old metal is granular in structure, containing minute holes which could collect ink and register roughly. This required further careful burnishing to push the copper back into a solid state again.

In normal circumstances a thin coat of steel would have been enough to protect the plates against damage or excessive wear during the repetitive printing, but early trials indicated that steel wore alarmingly after only a few sheets had rolled off. Something else was called for and chrome was the obvious answer because it is ten times harder than steel. It was another of Egerton-Williams' many contacts who came up with the solution; a friend who worked at the Royal Mint told him about chrome forming used in banknote production, where the master plate is copied in chrome and many duplicates are taken from it for the printing. The result is perfect notes and the method could be used for fine art printing if someone could be found to coat the *Florilegium* plates, but no other studio in London used this process and there were few businesses that had the expertise to handle the exacting job. Plenty of chroming shops existed, working on car bumpers and other commercial plating, but none was using the right kind of metal for printing purposes. Egerton-Williams was after hard chrome, the type used for instruments such as his burnishers, and in hydraulics. One firm did agree to tackle the task, but was declared bankrupt a few weeks before they were to start. Fortunately, two of their workers set up a new company and they were able to handle all the *Florilegium* chroming. In the meantime, Alecto decided temporarily to stop publishing contemporary artists in order to concentrate on *Banks' Florilegium*. The printing studios at 27 Kelso place were closed after 18 years of production and let to a design firm, although the company retained its offices there. A stock room and collating studio were acquired above the *Florilegium* premises at Appold Street, production began in earnest and the work continues at the time of writing.

The *Florilegium* staff are almost all from art schools. There is a small nucleus that has been with the studio from the start, including Mike Barratt, a qualified colour printer who was apprenticed at the age of 14 and is now the master printer who supervises the proofing. Martin Saw is also a master printer and Tom Milne is in charge of the hand-colouring. Their duties are interspersed with regular trips to the museum, but most of the editioning printers, working all day in the studio, usually leave after about a year because of the tedium of routine. There are no part-timers. Egerton-Williams concedes that the repetitive nature of the work, even in the most congenial surroundings, must lead to a certain drift: 'It doesn't matter what anybody says, it is one of the most boring things making print after print after print'. With this in mind he makes the ambience as attractive as possible and sets a good example by being on the premises longer than anyone else. 'If someone wants to work on a Saturday', he says, 'I'll be here. If they want anything to make the task easier, they can have it. We actually try to involve ourselves in being a proper company in preference to most etching studios which are run by a group of artists'. Egerton-Williams is an enthusiast for the latest advances in electronics and has installed an expensive sound system for use by the staff, along with wall-mounted television and excellent coffee available at all times. In his office, with its ever-open door and racks of fine wines, is a huge colour-coded chart on the wall to keep a check on print production and movements, and a computer to store colour information and other data about the entire *Florilegium* project.

'I have often seen this as a trial run for the studio we are going to have in the future', says Egerton-Williams. The project has kept 20 people employed at any one time and given good business to several contractors and suppliers. It has led to expansion of the studio and enhanced the reputation of its owner in the fiercely competitive world of fine arts printing. 'It's very good bread and butter', he admits, 'and when we're out on our own, there will be a facility that, unlike other companies, does not owe anybody a penny'. A small budget is set aside each year to allow experiments with new techniques and the latest printing technology: 'If the results are no good we chuck the equipment away, to the utter bewilderment of my accountant. We are trying to change the life of this type of studio'.

Joe Studholme acknowledges that on every level, the *Florilegium* has been the best thing his company has ever done, artistically, historically and financially: 'We are often asked how we could have spent such a long time on it, but I have never found it anything less than intriguing and satisfying'. The completion and placing of the edition with a full list of subscribers is justification enough for pride on the part of all who have contributed in any way to the realization of *Banks' Florilegium*. For all the principals it is proving a happy conclusion of a great project that had its origins in the eighteenth century; the British Museum (Natural History) has the satisfaction of seeing widespread distribution for one of its previously unpublishable treasures, Joe Studholme and Alecto have the artistic and financial kudos of being joint publisher with the museum of the world's largest fine arts direct printing, and Edward Egerton-Williams attained his ambition to own an etching studio and make it one of the finest establishments of its

Work in progress at the Egerton-Williams Studio.

kind. Their combined efforts have also focused attention on one of the most remarkable of Englishmen from the Age of Enlightenment. As Australians celebrate the bicentenary of European occupation of their southern continent, Joseph Banks' role in first suggesting the east coast of New Holland for British settlement may lead to a new assessment of his place in the

nation's history and perhaps the reinstatement of the title by which he was previously known: the Father of Australia.

In the meantime, Banks' engraved plates are left wearing their protective coats of chrome and returned to the museum in acid-free wrappers to rest, according to the agreement, for at least 50 years. Joe Studholme says: 'They will probably never be printed again because surely nobody would be so mad as to embark on producing 738 of them in colour. There's no point now they've been done, now they have been transformed from tarnished copper into nice colour prints'. There is, of course, no ritual defacement of the plates after the required number of copies have been pulled as is the fine arts convention. These engravings are icons of a commitment to furthering knowledge and, rather than being made redundant after printing, will continue to be treasured as irreplaceable objects, historically important to both Britain and Australia. They will, in the final analysis, have given delight to many more people than Joseph Banks could ever have envisaged, a great masterpiece of printing that finally met its deadline a couple of centuries late.

CHAPTER NOTES

1. The Voyage: London — Madeira 1768

EIGHTEENTH-CENTURY naval information came from the library at the National Maritime Museum, Greenwich, and most of my details about the *Endeavour* and preparations for her circumnavigation were gathered from this source. The Banks papers are dispersed to various libraries around the world. I have used material mainly from the National Library of Australia, Canberra, and Sydney's Mitchell Library to tell the story of Banks' younger years.

Joseph Banks and James Cook kept journals of the voyage and few incidents are left out of their daily entries. My narrative includes short excerpts from Banks' extensive writings to give an indication of his personality and a feeling for the language of the time. Cook's journal is more restrained, reflecting the rather taciturn nature of the man and his background. With both sources I have altered some spelling and applied punctuation to make for easier reading. As the journey progressed, the captain's output of words increased, often drawing heavily on Banks' entries after the two men compared notes. Logs were also kept by the vessel's lieutenants, recording information about weather, navigation and the routines of the ship. These were intended to be collected at the end of the voyage and sent to the Admiralty for record purposes. I made a research visit to Funchal, Madeira in September 1983.

2. Madeira — Rio de Janeiro 1768

THE narrative follows closely the journal descriptions of Joseph Banks, James Cook, Sydney Parkinson and an anonymous documentation believed to have been written by James Mario Matra. A comprehensive survey of what books were probably carried in the *Endeavour*'s library was made by D.J. Carr, Emeritus Professor of Developmental Biology at the Australian National University, Canberra. His article, '*The books that sailed with the Endeavour*' is published in *Endeavour*, New Series, Vol. 7 No. 4. 1983, London (Pergamon Press). A research visit was made to Rio de Janeiro in January 1984.

171

3. *Rio de Janeiro — Tierra del Fuego 1768–1769*

ONE of the best accounts of naval medicine and diet comes from the series of Maritime Monographs and Reports published by the National Maritime Museum, Greenwich. The one titled *Problems of medicine at sea* contains papers read at a symposium in March 1973 and includes a fascinating account of the changing character of sailors' diet and its influence on disease by Surgeon Commander G.J. Milton-Thompson.

Descriptions of shipboard life under sail in the region of Cape Horn are impressively described by the Australian-born writer and master of full-rigged ships, Alan Villiers. He travelled the Pacific in his vessel the *Joseph Conrad* following or crossing many of Cook's tracks. He describes Cook as the 'meticulous and infinitely careful explorer by sea, the most consistent and the greatest sailing-ship seaman there ever was'. Villiers wrote a dozen books and at least four of them are essential reading to capture the flavour of doubling the Horn and travelling the Pacific under sail: *Cruise of the Conrad*, *By way of Cape Horn*, *The way of a ship* and *Captain Cook, the seamen's seaman*.

4. *Tierra del Fuego — Tahiti 1769*

THE British voyages into the Pacific by Byron, Wallis and Carteret immediately preceding Cook's were documented by John Hawkesworth in the first part of his *An account of voyages . . .* published in London in 1773. As with the *Endeavour*'s journey, the treatment of the *Dolphin*'s two circumnavigations and that of the *Swallow* is none too reliable after Hawkesworth's literary filtering. Cook had copies of Byron's and Wallis' logs with him.

The Tahiti sojourn was the most extensively recorded part of the *Endeavour* voyage and the journal entries of Banks, Cook, Parkinson, Matra, Monkhouse, Molyneux, Pickersgill and others form the first sizeable literature in English about Polynesian life and customs. The impact of the noble savage on the eighteenth-century imagination was gathered from several sources including the popular writings of explorers, missionaries and novelists to the philosophic reasoning of Jean-Jacques Rousseau and the artistic background of Bernard Smith. The concept of a Pacific *beau sauvage* was largely illusory in the way that Utopia had been a figment of the Renaissance mind; both were science fiction to educated Europeans of the time. The papers from a symposium on the subject, *The opening of the Pacific — Image and reality*, held at the National Maritime Museum, Greenwich in November 1970 provide an interesting background. A research visit to Tahiti was made in January 1984.

5. *Tahiti — New Zealand 1769–1770*

ALMOST all my descriptions of the voyage around New Zealand are based on the journals of Joseph Banks and James Cook with occasional contributions by Sydney Parkinson and personal observation by the author.

By the second half of the eighteenth century, with techniques of navigation rapidly improving, some two-thirds of Australia was already outlined by Dutch navigators and the Englishman, Dampier. In addition, Tasman had named Van Dieman's Land (Tasmania) without knowing if it related to the rest of the continent. Cook studied all the information available to him about the mythical Terra Australis Incognita and it is unlikely that he would have confused New Holland with it. Bougainville, who was more than a year ahead of the *Endeavour*, was also seeking the missing continent and, after his sojourn in Tahiti, took his two ships west to discover Samoa and then arrive at the land Quiros had reached in 1606, to be known later as the New Hebrides (Vanuatu). At first Bougainville thought this was the eastern extremity of New Holland and to test his theory set sail due west. Nothing but open sea was encountered for the next week until they arrived off dangerous reefs with a continuous line of threatening breakers: Australia's Great Barrier Reef. If Bougainville's shipboard provisions had been more plentiful and if an entrance had presented itself through the reef, he would have become the first European known to have set foot on the east coast of New Holland, instead of Cook. The two French ships made haste to Batavia and arrived there in September 1768. After the *Endeavour* voyage Cook would continue to search in vain for the missing southern continent on his next journey of exploration in the Pacific.

6. New Zealand — Australia 1770

THE association of Cook, and to a much lesser degree Banks, with Australia provides some of the richest historical writing in the nation's brief history, and the material for this chapter is taken from their writings, together with visits to the sites of many of the *Endeavour* landfalls.

Apart from Botany Bay, in New South Wales, the other locations along the east coast where botanizing took place over a period of almost four months are all in Queensland: Bustard Bay, Quail Island, Thirsty Sound, Bay of Inlets, Palm Island, Rocky Point, Mission Bay, Cape Grafton, Point Lookout, Endeavour River, near Cape Flattery, Lizard Island, Cape Fear Islands, Eagle Island, Possession Island and Booby Island.

7. Australia — Java 1770

THE journals of Banks and Cook were the sources for this chapter, as well as a visit to Jakarta by the author in March 1984.

8. Java — London 1771

THE journals of Cook, Banks, Parkinson and Matra were the main sources for the material in this chapter. Cook's journal entries during this stage of the voyage, from Batavia, tell a grim tale of death. The first victim at sea was John Trusslove, the marine corporal, 'a man much esteemed by everyone on board' according to the captain. Cook was at a loss to explain the spread of infection through his restored and clean ship, reporting in his journal, 'Many of our people at this time lay dangerously ill with fevers and fluxes'. He attributed the sickness to water taken on at Princes Island, their final anchorage in Java, and soon had lime added to the casks for purification; later he ordered every accessible space between decks to be cleaned and washed down with vinegar in an attempt to stop the spread of infection. Banks' secretary Herman Spöring was the next to die, followed by marine private Thomas Dunster and then Sydney Parkinson. The names of the deceased mounted rapidly: soon after Parkinson it was the turn of the resilient old sailmaker, John Ravenhill, 'a man much advanced in years', according to Cook's journal entry, and then in the early morning of 29 January, the astronomer Charles Green met his end. After all his valued assistance with navigation across thousands of miles of uncharted ocean, he received at his demise another of Cook's admonitory epitaphs: 'He had long been in a state of bad health', the captain wrote, 'which he took no care to repair but on the contrary lived in such a manner as greatly promoted what he had long upon him, this brought on the flux which put a period to his life'. It was a harsh judgement of Green's drinking habits which may have weakened his resistance to disease, but dysentery struck indiscriminately, the strong, the weak, tippler and abstainer alike.

By the end of January, the previously extensive journal entries of both Cook and Banks languished into brief obituary notes and comments about wind and sea conditions; there was little else to note, and the simple act of wielding a pen used up too much energy. At the end of the month two of the carpenter's men died of the flux, Samuel Moody and Francis Haite, quickly followed by the cook, John Thompson, the seamen, James Nicholson and Archibald Wolfe, and the carpenter's mate Benjamin Jordan. On 2 February the gunner's servant Daniel Roberts died, on 3 February the sailmaker's assistant John Thurman, and on 4 February midshipman John Bootie and boatswain Jonathan Gathray were buried at sea. For those who were recovering their strength there was now some hope, although during the second week of February the capable midshipman Jonathan Monkhouse died as did John Satterley, the carpenter, 'a man much esteemed by me and every gentleman on board', according to Cook, who was alarmed that the carpenter's crew was now woefully depleted. Several of the seamen taken on board at Batavia also fell victim to dysentery, although the captain thought they might have been immune. The atmosphere surrounding so many burials at sea was constantly depressing and some of the men became highly agitated, regarding minor symptoms in themselves or their colleagues as instant death sentences.

Cook learned a little more about Bougainville's movements in the Pacific while the *Endeavour* rested at the Cape and they emphasized the political nature of his own voyage. In retrospect, the way the expedition had been organized probably indicated the concern of British authorities about French territorial ambitions. The *Endeavour* timetable is revealing. The Royal Society officially asked the king to finance an astronomical voyage on 15 February 1768, with no mention of exploration. Three weeks later King George signified 'his inclination to defray' the expenses. The Council of the Royal Society was told this on 25 March. By 1 April the Admiralty was advising the society of the purchase of a suitable vessel and asked 'to be informed of the number and quality of the persons intended to be sent to make the observations and what instructions are to be given the commander of the vessel'. This letter was read to the council on 3 April and the president reported he had recommended Alexander Dalrymple as commander, but that it violated navy rules, because he was a civilian. On 5 May the council learned that Cook was to command the ship and was proposed 'as a proper person to be one of the observers'. In the space of a mere seven weeks the king's agreement to defray expenses had been obtained, a vessel purchased and its commander appointed. Such matters, passing through several bureaucracies, did not usually move at this pace and suggest that the authorities viewed the Royal Society's request as a perfect opportunity to challenge French aspirations in the Pacific, and particularly the designs of Bougainville, under the guise of a scientific expedition controlled by the Admiralty.

9. Florilegium: Beginnings 1771–1820

THIS chapter spans the 49 years from the time of Joseph Banks' return on the *Endeavour* until his death. I have woven the story of work on the

Florilegium with his personal and professional activities where they either contribute to or hinder the progress of publication. As Banks became a celebrity and then a pillar of the establishment, the available documentation increases dramatically and I have drawn on Banks' papers from various sources, but principally in Canberra, Sydney, London and the United States. The most comprehensive information about the *Florilegium* is in a catalogue of the natural history collections of drawings, manuscripts and specimens made by Banks and Solander on the *Endeavour* voyage and held in the British Museum (Natural History). The aim of the three-part catalogue published by the museum is to relate all the drawings and manuscripts to the plant and animal specimens, including all the watercolour drawings, but listing only those specimens that are illustrated. Part 1 contains Botanical Collections: Australia, Part 2, Botanical Collections: Madeira, Brazil, Tierra del Fuego, Society Islands, New Zealand and Java together with a summary of the botanical collections, and Part 3 is devoted to zoology.

The other members of Banks' party on the *Resolution*, as well as Solander, were to be the fashionable painter, John Zoffany, three draughtsmen and nine servants, with accomplished musicians among them to serenade the gentlemen during the languors of the voyage. There were to be three astronomers on board including Dr James Lind from Edinburgh, who was also physician to the royal household at Windsor and a great lover of poetry and friend of Shelley.

By the time of the *Endeavour*'s return, botanical printing was becoming popular in continental Europe. The cult of the language of flowers gave rise to many popular books and fashionable young ladies eagerly took up flower painting. The Dutch-born Nikolaus von Jacquin had produced an exquisite work with his first publication in Paris in 1762. It was *Selectarum Stripium Americanarum Historia* and included nearly 200 of his drawings engraved under his own supervision and some of them hand-coloured. Later, Joseph Pierre Buc'hoz produced many folios of hand-coloured engravings including *Le Jardin d'Eden* in 1783 and *Le Grand Jardin de l'Univers* in 1785. In Vienna, Joseph Plenck's *Icones Plantarum Medicinalium* contained 738 hand-coloured engravings and was published between 1788 and 1812.

In England, Joseph Banks became familiar with the experiments being carried out by his genial painter friend, Paul Sandby, one of the foundation members of the Royal Academy, whose Italianate views of many parts of Britain and Ireland using gouache and watercolours were very popular. In 1768 Sandby had been appointed chief drawing master at the Royal Military Academy, Woolwich, a post he would hold for the next thirty years. Banks was one of Sandby's many patrons and travelled with him on a journey through Wales in a party that included Solander and Charles Greville. Greville had introduced the painter to the technique of aquatinting which obtained results similar to the broad flat tints of wash or ink drawings, but was formed instead by a minute web of lines on an engraved copper plate. The result of his experiments in this medium was the printing of the first important series of aquatint views in Britain, including scenes of Wales, and others of Windsor and Eton, and Warwick Castle. Paul Sandby also produced numerous engravings for publication and Banks was able to learn about the finer points of these methods from this distinctively English artist. There was also interest shown by Peter Perez Burdett in the use of colour for natural history illustrations, which he discussed with Banks, making particular reference to a volume of George Knorr's *Lapides diluvi universalis testes* of 1753 with its coloured plates. Burdett considered that whatever Banks had in mind, it would be an improvement in quality on that publication. A colour edition might have meant less engraving toil, but Banks was determined to get the best possible rendition by the ultimate subtleties of the engravers' art, so that every nuance of texture and tone could be expressed through the lines alone. After that was achieved there might be the opportunity to add some colour, but he always regarded the project as primarily scientific; any colour would be an artistic overlay to what had already been engraved in the copper.

Interest in botanical art had reached a high peak by the time of Solander's death. Its most famous practitioner was Pierre—Joseph Redouté who, in spite of his repellent appearance — but exquisite taste — attracted many of the most beautiful young ladies of Paris to his classes. Folios of his paintings in colour were on sale all over Europe in the late eighteenth century, produced from a technique known as stipple engraving which had developed in France and Holland. This was etching with dots rather than lines and it resulted in fine gradations of tone which were particularly appropriate for botanic illustration. Colour printing was carried out from a single plate by the application of the various inks with a little stub of fabric which happened to look like a child's rag doll — the technique is known as *à la poupée*. The plate was then re-inked for every impression taken from it. In claiming the invention of the process in 1798, Redouté boasted: 'We have thereby succeeded to our prints all the softness and brilliance of a water colour'. The English, for some insular reason, were not enthusiastic about stipple engraving, but on the continent it introduced a new style and delicacy to botanical art and Redouté, in particular, was made rich and famous by it.

Work probably began on the botanical drawings and engravings in preparation for publication during 1773. The artists engaged on the project were John Frederick Miller, his brother James Miller and, later, John Clevely, who together made 210 finished drawings from Parkinson's field sketches and notes. Thomas Burgis contributed three watercolours and the project was completed by Frederick Polydore Nodder with a personal total of 271 watercolours. The engravers included Daniel MacKenzie (251 plates), Gerald Sibelius (195) and Gabriel Smith (118), with a smaller output from Charles White, William Tringham, Robert Blyth, Frederick Polydore Nodder, Jabez Goldar, van Drazowa, Thomas Scratchley, John Lee, Jean-Baptiste Michell, William Smith, Edward Walker, John Roberts, Thomas Morris, Bannerman and Francis Chesham.

Banks' devotion to his work at Kew helped delay the *Florilegium*. Letters he wrote in 1791 and 1792, now in the British Library, reveal that his efforts at the Royal Botanic Garden were at the expense of his own publication. By the time he had lost all interest, Banks had spent more than £12,000 in preparing the fruits of the *Endeavour* voyage for printing, a sum approaching £1,000,000 in the currency of the mid-1980s.

Although Cook and Banks have received extensive attention, the third leading member of the voyage has largely been ignored. Immediately after the *Endeavour*'s arrival in Britain, Daniel Solander was the most famous of the trio and became a darling of society. Solander did not keep a journal and, therefore, his writings are less accessible. A valuable start to re-assessing his role in Banks' success and in botany generally, is a catalogue of his natural history letters and manuscripts in British collections by Judith A. Diment, Botany Department, and Alwyne Wheeler, Zoology Department, of the British Museum (Natural History). This extensive catalogue is published in *Archives of natural history* (1984).

It is thought that Banks' engravers pulled three sets of black ink impressions from the newly enraved plates as trial proofs and some of them were sent to other botanists. One set remains in the British Museum (Natural History), 28 plates make a folio volume in the Akademie Library, Berlin and others went to Stockholm, Paris, Berne, Leiden and possibly St. Petersburg, but how many were distributed is not known.

10. *The Long Wait 1820–1979*

DUPLICATE specimens from the *Endeavour* voyage were distributed by the British Museum (Natural History) in a diaspora of flora to various institutions. In the 1890s the National Museum, Wellington, New Zealand received 328 species and the Auckland Institute and Museum 249. The Royal Botanic Gardens in Sydney accepted 586 species of Australian plants in 1905. Other duplicates went to the Royal Botanic Gardens, Edinburgh; Botanischer Garten und Botanisches Museum, Berlin-Dahlem; Martin-Luther-Universität und Botanischer Garten, Halle; Dansk Botanisk Forening, Copenhagen; Muséum National d'Histoire Naturelle, Paris; Naturhistoriska riksmuseet, Stockholm; Naturhistorisches Museum, Vienna; Smithsonian Institution, Washington DC; Missouri Botanical Garden, St Louis; New York Botanical Garden; Central National Herbarium at Howrah, Calcutta.

The long and frustrating history of the Banks copper plate engravings from the time they were lodged in the British Museum in 1827 until the beginning of publication in full colour in 1980, is well documented in the introduction to the catalogue of the museum's natural history collections from the *Endeavour* voyage. My particular thanks go to the Botany Librarian, Judith A. Diment and her staff for their courtesies and guidance with information about the *Florilegium* and its many component parts.

The Alecto involvement in the project was recounted to me during several conversations with Joe Studholme in London, Plymouth, New York and Sydney, also taped interviews and by reference to many newspaper and magazine articles. A sizeable reference to print and electronic media items about the *Florilegium* will need to be compiled one day including articles, exhibitions, radio interviews and film, including the documentary made by Sailorman Films for screening at the 13th International Botanical Congress in Sydney in 1981 and the hour-long television feature, *The Flowering of the Pacific*, which I produced and directed to a Robert Hughes script for the co-production partners: the Australian Broadcasting Corporation, RM Arts, Munich, and Television New Zealand.

The appointment of Edward Egerton-Williams as the edition's master printer was discussed in taped conversations with Joe Studholme and Egerton-Williams during June 1984.

11. *Realization 1980*

THE studio procedures were explained during taped discussions recorded *in situ* with Edward Egerton-Williams and Joe Studholme in January 1984 and June 1984. Further information came from the studio staff, who were always courteous and co-operative in answering my many questions.

The principal colour technique adopted for the *Florilegium* was the

French *à la poupée* method, literally 'with a dolly', a little stub of tarlatan cloth used in the style of Johannes Tayler and Redouté to dab inks onto a plate. There was nothing particularly unusual about the procedure, although word got around that Egerton-Williams was reviving it after a disuse of many years. In fact, quite a number of sporting prints have been produced *à la poupée* and the French term is really just another way of describing multi-coloured single plate printing. Little research needed to be carried out into the actual production of the *Florilegium* because most of the techniques were well-known; the main consideration was to evolve the best way of organizing such a large edition within the bounds of financial possibility. 'What Edward is doing', Joe Studholme explained at the time, 'is a refinement of the old French technique but he carried out no actual studies, he just did it instinctively'.

After the copper plates are made ready for printing, there is the process of proofing. A black trial print is taken across town from the workaday East End to the fashionable environs of South Kensington where the proofing printer makes his notes while carefully inspecting the original watercolour sketch by Parkinson or one of Banks' London artists at the Natural History Museum. The originals are far too precious to leave the building and this makes it necessary for the Egerton-Williams' studio staff to make endless round journeys of ten miles in a constant shuttle appraisement. The need for regular cross-checking between the partners slows the whole procedure, but releasing the vulnerable originals and the obvious need for rigorous security at all times would prove to be equally inhibiting, and certainly more frustrating. Egerton-Williams thinks that better prints might have resulted from all the reference materials being under one roof, but there was no choice and he accepted these restrictions. Notes are made on the trial print about base colours and rubs for anthers, leaves, stamens and petals and what is called a 'dab sheet' is drawn up with all the colours needed for the engraving written on it. A special chart produced by the Royal Horticultural Society, which contains every nuance of botanical colour, is used by Alecto's printer and the museum's editor to identify the tones to be used at this critical juncture where science and art meet. The printer returns to the studio to mix up a little of each colour and ink areas of the plate for tests of sections of the image, each leaf, each frond, each flower, followed by the whole engraving. Then it is back to the museum to make further checks with Chris Humphries, one of the two editors of the edition. With complex images such as the New Zealand honeysuckle or Rewa-rewa (*Knightia excelsa*) or the many species of Banksia gathered by Banks and Solander at Botany Bay, it is often necessary for several visits before the colour registration is to everyone's satisfaction, and it can take up to two months to reach an acceptable result.

Egerton-Williams admits to more than an element of compromise because he must produce an image that looks good but is also practical enough to print many times over. He says: 'We can often make one that is bloody marvellous, but we could never get a hundred copies off it'. Alecto is seeking the best balance between the need for accuracy, in terms of the original watercolours, and the desire to show the beauty of the engraved lines, including their effects of tonal subtlety. Only rarely does Egerton-Williams have to consider the cost in time and, therefore, money of individual prints. He usually goes for the best result possible whether or not it is above a price that has been agreed with Alecto, although the collaboration of master printer and botanist is very effective in keeping to budget. Both Humphries and Egerton-Williams are pragmatic people with a common commitment, and their expertise is obvious in the quality of the issued prints. What is produced, however, is not always an exact version of the original finished sketch, even taking into account the subtle shifts of style and content during the many stages from living plant to completed print.

The compilation of information is supervised by the Natural History Museum with additional research supplied by Alecto's Elaine Shaughnessy. A list of plates accompanies each box of prints, and there is also the relevant botanical, geographical and historical information, together with the names of the artists and engravers included on each window mount in typography designed by Ian Mortimer. Judith Diment, the Botany Librarian, works closely with her co-editor on the taxonomy, which ensures the correct classification and name of each *Florilegium* specimen, particularly when Banks' or Solander's original nomenclature was incorrect. The watercolours have sometimes been found to contain errors in certain botanical details and there are a few plates which lack an original sketch as reference; sometimes only a black trial print survives or perhaps a pencil rendering. When this occurs it is the editors' responsibility to decide on the correct colours by deduction or sometimes by comparison with a living specimen. There is one plate with no colour references for a plant that has never been seen since and another, for a plant from Australia, where the watercolour is lost. In this latter case, Dr Donald McGillivray of the Royal Botanic Gardens in Sydney was asked to visit Botany Bay to compare the living flower with colour swatches supplied from London and send the results back to the museum for Chris Humphries to piece together, who then sent copies of his results back to Sydney for further checking. Small changes were made to the colour resolution and then, with all the information back in London, the printing could proceed. No one botanist can hope to be expert in all the world's flora and when Humphries is not certain of his information he sometimes makes reference to the experts at Kew Gardens, thus embracing another Banksian connection. The intention at proofing is to produce one reference print that is as near to perfect as possible. When this is achieved, it is embossed with the stamp of the British Museum (Natural History) and signed and dated by Chris Humphries. They call it a 'bat', a trade term derived from *bon attiré*, literally 'a good pull'. The specialist who will be in charge of a future stage of the process, the hand-colouring, is now brought in to see if there are any base colours to be used in the printing that need to be blocked out by colouring on top. White flowers, for instance, must be printed black so that they can be filled out by hand, leaving a fine black outline. If white ink was applied to the plate the flower would not register at all on the paper. Each of the 738 plates of the edition presents its own challenge.

Once a print is ready for editioning — that is for running the full number of copies — the inks are mixed for the entire run. They need to make 116 prints from the one engraved plate. In addition to the 100 numbered plates and the 10 *hors commerce*, there are three proof sets that, by tradition, go to the printers; another is split between those who repair the plates, proof and hand-colour them, and another goes, again by tradition, to the studio as a permanent record of the job. All the other proofing material will stay with the studio. A file is being compiled, which includes the bats and black copies, to form a comprehensive record in 64 books of every component of

179

the process. The finest pigments available from British suppliers are used for the inks and are supplemented with some special material from Charbonnel in Paris. Their mixing begins on glass sheets with a spatula. A ready-made tube could be bought for a few pounds but the studio's hand-mixed version will cost many times that amount because of its purity. No chalk is added to fill it out and, in any case, the particular colours used for the *Florilegium* are probably not available elsewhere. Sometimes as many as eighteen colours are prepared for one image and then stored in tubes ready for printing. At the end of proofing, the inks are ready, there is an approved reference print, and an offset print is taken from the plate on which is written all the information the editioning printer will need to know: the order the inks are to be applied and the precise areas where they must be placed. It is a sort of painting by numbers, filling in all the engraved lines on the copperplate with the right coloured ink, and wiping it away where it should print white. One person will now see the whole edition through to the end, totalling perhaps 130 separate prints because there is a normal rejection rate of about 10 per cent, although with some particularly complicated images as many as half might be winnowed.

The heavy grade paper is made to order by a mill in Somerset and, according to Edward Egerton-Williams, prints extremely well but, once again, the rejection rate is high: as many as 50 to 70 per cent of the sheets must be discarded, often for slight imperfections that only the closest scrutiny on a light box will reveal. This has something to do with Edward's personal quest for perfection and the commitment to producing better results than anyone else: 'Every print has to be the best we can do', he explains without apology, 'although we could let some inferior work through because one subscriber is unlikely to make a quality check with another'. In most studios it would just be work going out, but in this project he feels that his reputation rides on every image and even a minor flaw affects him like a deep wound. Since the project started no subscriber prints have been returned because of quality; in fact, the only serious query has concerned the few plates where an original engraving remained unfinished and that section of the print was rendered in a colour to look like a drawing, resulting in a few comments that the studio must have forgotten to complete its work. Attention to the smallest detail is one of the reasons why the paper has such a high rejection rate and a sheet is not used if there is the possibility that the resulting print will not be perfect. Edward explains: 'It's cheaper at the end of the day than discarding what's already been inked and printed, much cheaper to chuck out fifty sheets of paper at the beginning'.

Colour is applied with a variety of devices, not just the traditional *poupée* of the French colour printers. There is everything from rolled-up pieces of scrim to box after box of cotton buds which are available from the local chemist when supplies run low. The thin, stiff, cotton muslin, called tarlatan, used for the dollys, must be made to order because the material that is readily available sometimes contains minute specks of grit which could scratch the highly-polished surfaces of the plates. Tarlatan is also used for rubbing excess ink off the engraved surfaces, which wipe well because the chrome coating has a very low resistance, but additional hand cleaning is also done with chalk. Here again the emphasis is on meticulous detail. Most studios use powdered chalk for this work but Egerton-Williams insists on blocks to lift every vestige of grease from the plate and ensure that no dust is raised. A pure petroleum solution is also used for cleaning sections so that backgrounds intended to register white, print perfectly clear. Once the inking is completed the plate is placed in the bed of the hand press face up and dampened paper is placed on it face down with a sheet of tissue paper on top to stop any creasing and keep the back of the print clean. The paper has been dampened in a room off the main studio for precisely thirty minutes. If it is too wet there can be problems with blotching, and if on the dry side the image may be too faint. A thick blanket goes on top and then the editioning printer turns the large wheel which drives the bed beneath the rollers transferring with knife-edge pressure the inked image of the plate to the paper. When the pass has been completed, the blanket is raised, tissue paper taken away and the coloured print carefully peeled off. The procedure is almost exactly the same as in Banks' studio when his printers were taking their proofs — except for the application of colour.

From this point everything should be routine and the studio is structured so that, in theory, little is left to chance. But fine arts printing is an inexact business, as Edward is quick to explain: 'The process has so many variables: the thickness of the blanket, the roller pressure, the paper's dampness, the viscosity of the inks, the printer who's handling it all'. Without constant attention there could be different quality prints from the same plate, which are supposed to be identical. This means the editioning printer will need to be changing the blankets every so often, attending to the roller pressures and stripping the plate after a few impressions have been taken—only about 95 per cent of the inks are transferred to the paper and the residue remains lodged in the engraved lines, so after a short time they must be cleaned out with paint stripper before continuing the printing run. Edward devised a printing method for the *Florilegium* which relies on extreme pressure from the rollers on the presses more than in other studios, and this is combined with a heavier blanket than is usual. He could have settled for the common procedure of damp paper, runny inks and low roller pressures to give adequate results, but his method with partly damp paper, thick ink and increased pressures results in a superior, very precise line. It has meant, however, that some of the rollers on the Rochat presses have cracked under the strain, which has not happened anywhere else before, and this on some of the strongest machinery available.

The prints are left to dry and the next day they are placed between blotters: print, blotter, print, blotter ... until after six prints and seven blotters a wooden board of the same dimensions is placed on top to level the stack. Then the pile grows twelve times in the same proportions of prints, blotters and boards until finally weights totalling 200 kilograms are placed on top. The stack remains that way for a day before being rebuilt with fresh blotting paper and then left for a week, by the end of which the prints are perfectly dry. Checking comes later and it is almost certain there will be some unwanted marks that have been picked up along the way. These are then individually cleaned and re-soaked because in the cleaning the paper might pick up a few extra scuffs. The blotting sequence is gone through again and, at the end, perfect prints should emerge ready to be hand-coloured under the controlled studio lighting, the final stage of their production.

The colouring can take as little as ten minutes for each image, or the addition of painstakingly laborious detail might see only three prints completed by one person in a full day. The *Florilegium* began by using a maximum of six or seven colours to be applied by hand, but as techniques

181

were refined, up to 20 colours were used. Hand-colouring developed into what Edward describes as 'a full-blooded assault on certain areas'. The main attention is given to flowers and the more delicate botanical details, generally leaving the larger leaf areas to be expressed through the engraved line and the colour printed on the press. Many of the flowers are so delicately engraved that colour cannot be applied to the plate because the smallest possible application is from a cotton bud with a width of 1/8 of an inch, anything smaller cannot be inked. It is therefore left until the hand-colouring stage when fine brushes can add the precise colours required. After this is completed the prints are again checked carefully by the same person who inspected them when they first came from the press. If everything is perfect, they will be stamped, endorsed with print number, plate identification, edition number and printer's name. They are then ready to be boxed and sent out to subscribers.

The prospectus for *Banks' Florilegium* issued by the partners in 1980 stated the intention of the publication:

'*Banks' Florilegium* is published now in the firm belief that from the combined points of view of science, history and the art of botanical engraving there is no satisfactory substitute for a comprehensive printing from the original plates. The historical interest and aesthetic quality of these engravings speak for themselves. From the scientific point of view the engravings are highly relevant to the correct application of a number of botanical names. They have the advantage of depicting species in a living state. These species were later named and described, often inadequately, from the dried specimens. *Banks' Florilegium* will facilitate comparisons between the earliest graphic depictions and subsequent written descriptions'.

In terms of the printing method chosen, the prospectus states:

'The present publication will, for the first time, be printed in colour *à la poupée*. The Publishers decided to use this technique (developed by the Dutchman Johannes Tayler at the end of the seventeenth century) after experiments demonstrated that it produces the most satisfying overall result scientifically and aesthetically. It achieves the best balance between the need to be accurate in terms of the original water-colours and the need to show the beauty of the engraved lines including their effects on tonal subtelty.'

When work on the *Florilegium* became virtually a routine operation, Alecto took on additional assignments, including a limited edition of large-format prints for the American Museum of Natural History in New York, of six of their prized Audubon plates of American birds, followed by a facsimile printing of the *Domesday Book* to mark its 900th anniversary in 1986 and then a large suite of graphic works by Sir Sidney Nolan on themes relevant to Australia's bicentenary in 1988.

BIBLIOGRAPHY

THIS selected list includes complete publications relating mainly to the *Endeavour* voyage, Cook and Banks, the scientific results of the voyage, and Editions Alecto.

BADGER, G.M. (editor), *Captain Cook, navigator and scientist*, Australian National University Press, Canberra, 1970

BAGLIN, D. AND MULLINS, B., *Captain Cook's Australia*, Horwitz, Sydney, 1969

BANKS, J., *The Journal of Joseph Banks in the* Endeavour, Genesis Publications, Guildford, 1980

BEAGLEHOLE, J.C., *The exploration of the Pacific*, A.C. Black, London, 1934

—— (editor), *The journals of Captain James Cook on his voyages of discovery*. Vol. 1 *The voyage of the* Endeavour *1768–1771*, Cambridge University Press for the Hakluyt Society, 1955

—— (editor), *The* Endeavour *journal of Joseph Banks 1768–1771*, The Trustees of the Public Library of New South Wales in association with Angus & Robertson, Sydney, 1962

—— *The life of Captain James Cook*, A.C. Black, London, 1974

BEGG, A.C. and BEGG, N.C., *James Cook and New Zealand*, A.R. Shearer, Government Printer, Wellington, 1969

BLUNT, W., *The art of botanical illustration*, Collins, London, 1950

BLUNT, W. and STEARN, W.T. (editors), *Captain Cook's Florilegium*, Lion and Unicorn Press, London, 1973

BRITTEN, J., *Illustrations of Australian plants collected in 1770 during Captain Cook's voyage round the world in HMS* Endeavour *by the Right Hon. Sir Joseph Banks ... and Dr Daniel Solander ... with determinations by James Britten*, British Museum (Natural History), London, 1900–1905

BRYCE, G., *The sketchbook of HMS* Endeavour, Collins, Sydney, 1983

CAMERON, H.C., *Sir Joseph Banks*, Batchworth, London, 1952

CAMERON, R.W., *The golden haze: with Captain Cook in the Pacific*, Weidenfeld and Nicolson, London, 1964

CARR, D.J. (editor), *Sydney Parkinson*, British Museum (Natural History) in association with Australian National University Press, Canberra, 1983

CARRINGTON, H., *Life of Captain Cook*, Sidgwick and Jackson, London, 1939

CARTER, H.B., *Sir Joseph Banks*, British Museum (Natural History), London, 1985

CLAIR, C., *Captain James Cook the navigator*, Bruce and Gawthorn, London, 1963

COOK, J., *Captain Cook: his artists, his voyages*, Daily Telegraph, Sydney, 1970

—— *The journal of HMS* Endeavour *1768–1771*, Genesis Publications, Guildford, 1977

DIMENT, J.A. and HUMPHRIES, C.J. (editors), *Banks' Florilegium*, Alecto Historical Editions in association with British Museum (Natural History), London, 1980–

—— (editors), *Catalogue* Endeavour *Voyage* (3 parts), Archives of Natural History, British Museum (Natural History), London, 1984–

FISHER, R. AND JOHNSTONE, H. (editors), *Captain James Cook and his times*, University of Washington Press, Seattle, 1979

GREENHILL, B., *James Cook: the opening of the Pacific*, Her Majesty's Stationery Office, London, 1970

GWYTHER, J., *Captain Cook and the South Pacific: the voyage of the* Endeavour *1768–1771*, Houghton Mifflin, Boston, 1954

HAWKESWORTH, J., *An account of voyages undertaken by the order of His present Majesty ... Drawn up from the journals which were kept by the several commanders, and from the papers of Joseph Banks esq.*, London, 1773

HOLMES, M., *Captain James Cook, R.N., F.R.S.: a bibliographical excursion*, Francis Edwards, London, 1952

HOOKER, J.D., *Journal of the Right Hon. Sir Joseph Banks ...* Macmillan, London, 1896

JOPPIEN, R. and SMITH, B., *The art of Captain Cook's voyages*, Vol. 1 *The voyage of The Endeavour*, Oxford University Press, 1984

KIPPIS, A., *The life of Captain James Cook*, London, 1788

KITSON, A., *The life of Captain James Cook, the circumnavigator*, Murray, London, 1911

LEE, I., *Early explorers in Australia*, Methuen, London, 1925

LEMMON, K., *The golden age of plant hunters*, Phoenix House, London, 1968

LLOYD, C., *The voyages of Captain James Cook*, Cresset Press, London, 1949

LYSAGHT, A.M., *Joseph Banks in Newfoundland and Labrador. 1766*, Faber & Faber, London, 1971

LYTE, C., *Sir Joseph Banks*, Reed, Sydney, 1980

McCORMICK, E.H., *Omai: Pacific envoy*, Auckland University Press, 1977

McLEAN, A., *Captain Cook*, Collins, London, 1972

MACKANESS, G., *Sir Joseph Banks: his relations with Australia*, Angus & Robertson, Sydney, 1936

MAIDEN, J.H., *Sir Joseph Banks: the 'Father of Australia'*, William Applegate Gullick, Government Printer, Sydney, 1909

MATRA, J.M. (attrib.), *A journal of a voyage round the world in His Majesty's Ship Endeavour ...* London, 1771

MEGAW, J.V.S. (editor), *Employ'd as a discoverer*, Reed, Sydney, 1971

MILLER, E., *That noble cabinet*, Andre Deutsch, London, 1973

MOOREHEAD, A., *The fatal impact*, Hamilton, London, 1966

MORRELL, W.P. (editor), *Sir Joseph Banks in New Zealand, from his journal*, Reed, Wellington, 1969

MUIR, J.R., *The life and achievements of Captain James Cook*, Blackie, London, 1939

PAGET, H., *To the south there is a great land: Captain Cook, Sir Joseph Banks and Australia*, The Australian, Sydney, 1970

PARKINSON, S., *A journal of a voyage to the South Seas in His Majesty's Ship the* Endeavour, London, 1773

PRICE, A.G., *The exploration of Captain James Cook in the Pacific (1768–1779)*, Angus & Robertson, Sydney, 1969

RIENITS, R. and RIENITS, T., *The voyages of Captain Cook*, Hamlyn, London, 1968

ROBERTSON, J., *The Captain Cook myth*, Angus & Robertson, Sydney, 1981

SMITH, B., *European vision and the South Pacific*, Clarendon Press, Oxford, 1960

SMITH, E., *The life of Sir Joseph Banks*, John Lane, Bodley Head, London, 1911

SPENCER, C. (editor), *A decade of printmaking*, Academy Editions, London, 1973

SUTTOR, G., *Memoirs historical and scientific of ... Sir Joseph Banks*, E. Mason, Sydney, 1855

TOMLINSON, C., *Sir Joseph Banks and the Royal Society*, London, 1844

VILLIERS, A., *Captain Cook, the seamen's seaman*, Hodder and Stoughton, London, 1967

WHARTON, W.J.L. (editor), *Captain Cook's journal during his first voyage round the world made in H.M. Bark* Endeavour *1768–1771*, London, 1893

WILLIAMSON, J.A., *Cook and the opening of the Pacific*, Hodder and Stoughton, London, 1946

INFORMATION CONCERNING
BANKS' FLORILEGIUM

THE work, published by Alecto Historical Editions in association with the British Museum (Natural History), is produced in 34 parts:

I–XV Australia 337 plates
XVI Brazil 23 plates
XVII–XVIII Java 30 plates
XIX Madeira 11 plates
XX–XXVII New Zealand 183 plates
XXVIII–XXXI The Society Islands 89 plates
XXXII–XXXIV Tierra del Fuego 65 plates

A total of 738 plates. The publishers have followed the divisions established by the Botany Library in the British Museum (Natural History) where the collection is arranged geographically into 18 volumes. These contain Sydney Parkinson's original sketches alongside the finished watercolour drawings, the black and white proofs of the line engravings taken in the eighteenth century and, in the case of most but not all of the Australian plants, the photo-lithographic reproductions from James Britten's *Illustrations of Australian plants collected in 1770*. The museum's herbarium has examples of most of the plant specimens collected on the *Endeavour*.

Editors: J.A. Diment and C.A. Humphries, British Museum (Natural History).
Master printer: Edward Egerton-Williams at Egerton-Williams Studio.
Method: Colour printing *à la poupée*, colours worked directly into the single plate before each print is pulled, with additional watercolour touches.
Plate size: Virtually uniform: 18 × 12 inches (457 × 305 mm).
Paper: Somerset mould-made 300 gsm, each sheet watermarked 'AHE', produced in a special making by the Inveresk Paper Company: $28\frac{1}{2}$ × $21\frac{7}{8}$ inches (724 × 556 mm). Light cream printing, textured surface.
Presses: Manufactured by Harry F. Rochat, Barnet.
Pigments: Pure pigments from various British sources and F. Charbonnel, Paris.
Botany text: Compilation of botanical information under the direction of British Museum (Natural History), consisting of lists of plates to accompany each part, and of the relevant botanical, geographical and historical facts (including names of artists and engravers) printed on individual mounts. Research for Alecto Historical Editions by Miss E. Shaughnessy.
Typography: Designed by Ian Mortimer, set and printed on hand presses at I.M. Imprint.
Typefaces: Founder's Caslon Old Face and Old Face Open.
Mounts: Each of the 738 engravings protected within a double-fold sheet of Somerset mould-made 300 gsm paper, acid free and cut to form a window mount.
Solander cases: Each of the 34 parts presented in its own Solander case made by G. Ryder & Company Limited.
Edition: Limited to 100 complete sets numbered in pencil on the verso '1/100 to 100/100, with an additional ten sets *hors commerce* numbered 'I/X to X/X'. Each sheet stamped with the publishers' marks.

Institutions subscribing to *Banks' Florilegium*

	SET NUMBER
United Kingdom	
British Museum (Natural History)	HC9 and HC10
Royal Society	39/100
University Library, Cambridge	47/100
Central Library, Belfast	65/100
City Library, Birmingham	64/100
National Library of Scotland	66/100
National Library of Wales	75/100
France	
Muséum National d'Histoire Naturelle, Paris	93/100
Germany	
Botanischer Garten und Museum, Berlin	41/100
Sweden	
Royal Academy of Sciences, Stockholm	69/100
University of Uppsala	48/100
Australia	
National Library of Australia, Canberra	32/100
State Library of South Australia, Adelaide	12/100
State Library of NSW, Sydney	13/100
State Library of Queensland, Brisbane	17/100
State Library of Tasmania	14/100
Art Gallery of Western Australia, Perth	24/100
Royal Botanic Gardens, Melbourne	11/100
Reid Library, University of Western Australia, Perth	37/100
University of Sydney	3/100
New Zealand	
Alexander Turnbull Library, Wellington	40/100
Canada	
National Library, Ottawa	91/100
United States	
Humanities Research Center, Texas	22/100
Missouri Botanical Garden, St Louis	81/100
*Yale University Library	82/100
*Dallas Botanical Garden	83/100
Ohio State University, Columbus	85/100
Louisiana State University, Baton Rouge	43/100
New York Public Library	87/100
*Pierpont Morgan Library, New York	21/100
*Linda Hall Library, Kansas City	86/100
*Spencer Research Library, University of Kansas	68/100
Filoli Center, California	96/100
*University of California, Davis	77/100
University of California, Los Angeles	90/100

Given by, or on long-term loan from a private collector.

ILLUSTRATIVE
ACKNOWLEDGEMENTS

Prints from *Banks' Florilegium* by permission of British Museum (Natural History) and Alecto Historical Editions

Page 4   National Library of Australia
Page 8   National Library of Australia
Page 12   Parham Park, Sussex
Page 13   *Left:* Brabourne Collection
          Centre: Parham Park, Sussex
          Right: British Museum (Natural History)
Page 20   British Museum (Natural History)
Page 32   British Museum (Natural History)
Page 37   British Museum (Natural History)
Page 64   British Museum (Natural History)
Page 78   British Museum (Natural History)
Page 95   National Library of Australia
Page 98   British Museum (Natural History)
Page 99   British Museum (Natural History)
Page 102 British Museum (Natural History)
Page 111 British Museum (Natural History)
Page 130 National Library of Australia
Page 131 Parham Park, Sussex
Page 137 National Library of Australia
Page 138 National Library of Australia
Page 139 Parham Park, Sussex
Page 146 Brabourne collection
Page 147 *Left:* Brabourne collection
          Right: Brabourne collection
Page 148 British Museum (Natural History)
Page 149 National Portrait Gallery, London
Page 150 National Library of Australia
Page 151 Parham Park, Sussex
Page 154 National Portrait Gallery, London

INDEX